BLOWN to BITS in the MINE

A History of Mining & Explosives In the United States

Eric Twitty

WESTERN REFLECTIONS
PUBLISHING COMPANY®

Lake City, Colorado

ISBN: 978-1-932738-82-7
Library of Congress Control Number: 00-105541

Cover design: Angela Hollingsworth, APH Creative Design

Printed in the United States of America

Western Reflections Publishing Company
P. O. Box 1149
951 N. Highway 149
Lake City, Colorado 81235

www.westernreflectionspub.com
westref@montrose.net

TABLE OF CONTENTS

ACKNOWLEDGMENTS

I am indebted to a number of people and several institutions who were generous with their time and materials. Without their help, this work would not have been possible. First, I give a hearty thanks to the Hagley Museum and Library in Wilmington, Delaware, for opening their book stacks, photo archives, manuscripts, and artifact collection to me during my stay. I would like to single out Robert Howard, curator, for sharing his personal files and research data, his vast knowledge, and the time to assist and guide parts of my research on explosives.

Other institutions which contributed significantly are E.I. DuPont de Nemours & Co., Hercules Inc., ICI—Atlas, and the Institute of Makers of Explosives. They generously allowed me to use their material for illustrations.

Without the assistance from the community of people dedicated toward the preservation and interpretation of historic mines and their artifacts, my archaeological and artifact analysis would not have been possible. I especially thank Lane Griffin, Andy Martin, and Brian Schrage. Thanks also to other people who opened their collections to me at inconvenience to themselves, especially J. Scott Altenbach in Albuquerque, New Mexico; Lane Griffin in Reno, Nevada; Andy Martin in Tucson, Arizona; Leo Stambaugh in Colorado; Bob Hauck and the Sterling Hill Mining Museum in Ogdensburg, New Jersey; Steve Kohler in California; Kurt Kremer in Pennsylvania; Larry Kuester in California; Frank Schiavo and the Western Mining & Railroad Museum in Helper, Utah; Bill Lorah in Pennsylvania; and the World Museum of Mining in Butte, Montana. All of the above went out of their ways to help me in collecting data. Also of note are John Kynor, Deric English, Dan Lockhart, John Pollowski, Bob Otto, Dale McNee, Bob Everleth, Mark Bohannan, and Dave and Don White.

I certainly am grateful to those who worked in the mines and took the time to chat with me. The most prominent are Ed Hunter in Victor, Colorado; Ray Sporr in Delta, Utah; Walter Johnson in Ely, Nevada; and Phillip Polly in Luning, Nevada. Thanks to all for facilitating the preservation of the art of blasting and mining.

INTRODUCTION: THE STORY OF MINING & BLASTING

The introduction of blasting technology to North American mines and stone quarries set in motion a chain of events which propelled the Industrial Revolution. The art of blasting made possible mining, quarrying, and the development of a transportation system on the immense scale capable of meeting the demands of the Industrial Revolution in North America. Blasting permitted mining companies to tap into North America's vast pool of precious and industrial metals, giving the United States the economic and material wealth that transformed the nation into an industrial juggernaut.

Blasting technology reduced the costs of mining while increasing output, and as a result mining in North America skyrocketed beginning in the 1850s, stimulating social, political, and economic changes that collided in a tangled web. Because of reduced costs and greater production, mining fell within economic and productive reach of private capitalists and small companies. Seeking to profit from precious metals primarily, and industrial metals secondarily, capitalists, small mining companies, and prospectors armed with blasting technology then set out to comb North America for riches, drawing Euro-American settlement behind. The boom of mining and associated ore milling created a huge pool of jobs requiring both highly skilled and unskilled workers, enticing immigrants from Europe and Mexico. Mining, dependent on easily depleted mineral resources, fostered a subclass of workers who needed to be much more mobile than their manufacturing-industry and agricultural counterparts. As a result, the spread of Euro-Americans across North America and distribution of wealth among them created significant social and political change, and presented both Euro-American and Native American societies with major problems.

On an equally important scale, the use of explosives in mining created a unique and potentially deadly work environment for miners. Explosives presented miners with a wide array of problems ranging from chronic exposure to poisonous byproduct gases, suffocation in totally fouled mine atmospheres, injury and death from accidental explosions, and injury from drilling equipment. However, considering how brutal mining was without explosives, and the difficulty of pitting human muscle and bone against mother rock, miners indeed heavily benefited from blasting. In place of attempting to break rock by striking it with steel hammers and bars, miners used at first hand-drills, and by the 1910s machine-drills, to bore holes into which they inserted explosive charges. Employing the practice of blasting, which consumed most of a miner's eight-to-ten hour shift, was much easier and less arduous than mining without explosives.

Between the 1830s and 1850s explosives became a permanent feature of the miner's tool chest; miners and engineers alike saw explosives as essential for mining as pick, shovel, and illumination. Miners quickly learned how to handle blasting powder, the only explosive used in mines until the introduction of nitroglycerine during the 1860s, primarily from English coal and Cornish metal miners. Drilling and blasting methods based on powder were straightforward and labor-intensive, and American miners became expert at the process. However, when industrialists introduced dynamite and the mechanical rockdrill to North American mines in the 1870s, American miners began the ascension of a steep learning curve punctuated by close calls, injuries, and deaths. Causes for accidents proved to lie with the miner, the mining company, and the mine as a work environment. Many accidents were caused by miners acting foolishly out of haste, laziness, and in some cases out of fear of losing their jobs. Many mining companies, guilty of fostering a poor and unsafe work culture, exacerbated the potential for accidents by pushing miners for tonnage while minimizing safety expenditures, in terms of both materials and time. Significant reduction of accidents, which occurred beginning in the 1910s, came about primarily by miners learning how to handle explosives safely, through changes in the corporate culture of mining companies, and by reformulating explosive products. Contrary to mining during the Gilded Age, by the 1930s coal and metal mining companies alike were competing for records of the greatest number of accident-free mandays. Mirroring a nationwide shift in corporate, government, and social attitudes toward labor in North America during the first half of the twentieth century, the mine as a work environment, and its effect on the lives and well-being of miners, became a focal point of improvement. Finally, by the 1930s headlines with titles such as "Blown to Bits in the Mine" in newspapers and reports had, mercifully, become a noteworthy rarity.

CHAPTER ONE

EXPLOSIVES USED FOR MINING

Among all of the contraptions, inventions, and machines used in mines throughout North America, explosives had perhaps the greatest impact on the lives of hardrock and coal miners. Explosives' byproduct gases fouled mine atmospheres, and their temperamental nature required constant vigilance, which if relaxed, could have resulted in tragedy. By the 1840s the practices of blasting had become synonymous with American mining, and explosives held an omnipresence in the mine workplace.

At first the only explosive for blasting available to American miners was blasting powder, also known as black powder. Then came dynamite in the late 1860s. From the moment of its introduction to North America, dynamite proved to be much better than blasting powder in hard rock, and in a little more than ten years won its way into hardrock mines throughout the continent. Most metaliferous ores occurred in hard rock, while coal and nonmetaliferous minerals tended to be deposited in soft rock. Dynamite's superior performance in hard rock drove metal mining companies to quickly shun blasting powder.

For the first two decades of dynamite's existence, hardrock miners learned through accidents, brushes with death, and tragedies its behavioral characteristics and its advantages as a superior blasting agent. The learning curve experienced by miners was steep in the beginning, it began to flatten out by the 1900s, and by the 1920s explosives-related accidents were a rarity.

Like most forms of mining technology, explosives underwent improvement, and the overall product line expanded to serve mining's needs. During the 1870s explosives makers offered the mining and quarrying industries primarily blasting powder and straight dynamite. During the 1880s and 1890s the major explosives manufacturers expanded their product lines to include gelatin, extra dynamite, extra gelatin, ammonium nitrate, and non-nitroglycerine dynamites. Most of these new explosives benefited hardrock miners because they produced less poisonous gases and they were safer to handle, but miners adopted them only slowly.

As much as explosive products meant to miners as a work group, they also held considerable significance for mining as an industry. Dynamite and its cousins forwarded hardrock mining, increasing productivity, speeding development of underground workings, and decreasing production costs. A drop in the price of bulk explosives around the turn of the century made a new form of mining possible—open pitting. This type of mining was by rule explosives-intensive because thousands of tons of rock had to be blasted and removed at a time to achieve ore production in economies of scale. By the 1920s the open pit mining industry became a significant explosives consumer in and of itself, stimulating explosives companies to develop specialty products designed to move huge quantities of rock at little cost.

The evolution of explosives also improved the state of coal mining. For decades, coal miners used blasting powder almost exclusively because it cost least, and it performed best in coal. Beginning in the 1890s, explosives manufacturers attempted to adapt dynamite to meet the requirements of blasting the relatively soft fossil fuel, and they reformulated it for a lower explosion temperature to minimize the danger of igniting methane and coal dust.

The story behind the development of explosives is a fundamental part of mining and blasting in North America. The different types of explosives presented both costs and benefits to miners and mining companies, which all too often were at odds. Explosives were as common a tool for moving rock as the shovel, and in this chapter the miners' tool chest is laid open.

Blasting Powder: A Cornerstone of the Industrial Revolution

Blasting powder, the earliest and for some time North America's most popular explosive, played a fundamental role in both mining and the Industrial Revolution. It alone made possible mining the metals and fuels needed for heating, smelting, and manufacturing on the scale demanded by the Industrial Revolution. Like all explosive products used for mining, blasting powder experienced a process of development, evolution, and specialization. From the first attempts at blasting in the early 1600s until the early 1800s, those miners daring enough to blast did so with gunpowder, blasting powder's predecessor. Gunpowder also underwent improvement, and when manufacturers began to improve their gunpowders in the early 1800s, it became too expensive for mining, and the product's performance became specialized. This marked the rise of blasting powder, based on an early gunpowder formula, as a blasting-specific agent. The advantages possessed by blasting powder lay in simple, abundant ingredients, and relative ease of manufacture.

Explosives manufacturers made blasting powder by combining finely powdered sulfur, charcoal, and saltpeter in a specific ratio. The ingredients of the quickest-burning blasting powders, used in hard rock and in stone quarries, were mixed in ratios 1 : 1 : 6 and 1.2 : 0.8 : 6. Manufacturers used variances such as 1.5 : 1.3 : 7.2 and 3 : 3 : 4 as a means of adjusting for impurities and maintaining performance while reducing expensive saltpeter[1]. These variances also slowed the explosion into a heaving and moving force, which was best for blasting soft rock.

Blasting powder combusted in an oxidation reaction that produced new compounds, most of which were byproduct gases consisting of mostly carbon dioxide and nitrogen, as well as traces of carbon monoxide, hydrogen sulfide, hydrogen, and methane, and solids. When damp powder combusted, the composition of the above gases changed to include poisonous gases typified by an excess of carbon monoxide, hydrogen sulfide, and the evolution of poisonous carbonic oxide[2]. These byproduct gases proved themselves to be of great significance to miners because they were omnipresent in mine atmospheres. On average, blasting powder's gases occupied approximately 300-400 times the powder's original volume. The temperature of a complete explosion when contained in a drill-hole averaged between 2,000° F and 4,000° F.

Mining engineers classified blasting powder as a *low explosive*, defined as a substance that exploded only when sufficiently confined; when unconfined the powder merely burned[3]. An explosion occurred only when gases created during combustion were blocked, which for blasting in mines, was achieved by placing the powder in drill-holes.

Because many different rock conditions were encountered in mines, adapting explosions to the different rock types was important for the most efficient use of blasting powder. A quick, shattering explosion was not good in soft and crumbly rock such as coal and shale because energy was wasted pulverizing the material into undesirable, small fragments without moving the mass much. However, a quick, shattering explosion was very effective in hard rock because that kind of energy was most effective for fracturing it. To control the speed of explosion, miners and quarrymen used powder that manufacturers had sorted into uniform grain sizes. Fine grains presented greater surface area, allowing for rapid combustion and, most importantly, for rapid escape of gases[4]. When confined, this resulted in a quicker and more shattering explosion. Large grains produced a slower explosion because they offered less overall surface area. To denote grain sizes powder manufacturers used a scale with *FFFFFF* as the finest, *F* as medium-small, *C* as medium large, and *CCC* as the most coarse. Until the 1890s there was some variation of grain sizes among manufacturers within this designation. Gunpowder had similar labeling, but manufacturers used *FG*, *FFG*, and so on, although some manufacturers prior to the 1910s occasionally applied this designation to their blasting powders. All powder makers labeled their kegs with one of these terms.

The Beginning of Mining and Blasting

Blasting powder played a pivotal role in the general history of blasting and mining. It was the first explosive ever used for blasting, and it was the only explosive American miners would know for over 100 years. The history of blasting began in 1627 at a mine in Shemnitz, Hungary, where Kaspar Weindl documented the first use of gunpowder to fracture rock[5]. This first blast was a simple affair—a mining engineer stuffed gunpowder into some crevices in a tunnel's working face as an experiment, and ignited the explosive with a primitive fuse. The shot, although undoubtedly only marginally effective, proved reasonably successful. Excited at the prospects of using something other than brutal labor to break rock, mining engineers conducted further experiments and spread the concept of blasting through Northern Europe. Engineers subsequently blasted in mines at Clausthal, Germany, in 1632, at Freiberg, Germany, in 1645, and in construction of the Albula Road in the Swiss Alps in the 1690s. As wonderful as blasting was, the practice spread slowly because drilling holes into which miners loaded powder was an unsure practice, and the quantity of gunpowder needed was costly.

EXACT SIZE OF GRAINS OF "A" AND "B" DU PONT BLASTING POWDERS

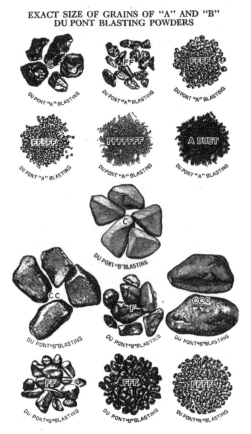

Figure 1. The chart illustrates actual grain sizes and shapes for types A and B blasting powders. Grains FF and smaller combusted quickest and were best for blasting hard rock, while grains F and larger combusted slower and were suited for soft material such as coal. (Source: E.I. DuPont de Nemours & Co., 1932, p 8.)

Once German mining engineers, recognized as the best in their field and in demand throughout Europe, developed primitive drilling technology and methods of loading and firing explosives, they educated the rest of Europe and Great Britain in the ways of breaking rock with powder. German engineers brought blasting to Cornwall in 1670 and, under the employment of Spanish mining interests, introduced it to North America in Mexico during the eighteenth century[6]. Over the course of several generations the Cornish miners significantly improved the then-primitive blasting methods, and they brought their practices to North American mines between the 1820s and 1850s.

The 1830s was one of the most pivotal decades the mining industry would undergo, because entrepreneurs devised three inventions which revolutionized the technology of blasting, and also stimulated a voracious appetite for coal and metal ores. First, American steel makers began using anthracite coal for smelting iron ore because it was far superior to the hardwood charcoal used until then. Around the same time an American inventor patented a grate which greatly improved the efficiency of burning lump-coal in fireplaces. These two inventions caused the demand for coal and iron ore to soar. The skyrocketing demand created a heavy interest in mining, and it was the practice of blasting that allowed fuel and metal mines to be worked on the necessary industrial scale.

The third and most significant invention was William Bickford's creation of *Miner's Safety Fuse* in 1831 in Cornwall, England[7]. Bickford, who was very sympathetic to the hard life of Cornwall's miners, hit upon a bright idea while watching a hemp rope-making machine. Why not adapt a special funnel through which high-grade gunpowder could be fed and have the machine braid its twine around the powder? After a little development, William Bickford came up with a device which wound several layers of twine around a powder core, with the threads of each successive layer being wound at an opposite 90 degrees. The final product was absolutely ingenious! Known as *Bickford's Fuse, Countered Safety Fuse*, and *Safety Fuse*, it was an endless textile tube approximately one quarter-inch in diameter with an inner powder train. The value of safety fuse lay in its predictable burning rate, the protection its varnished sheath offered against water, and the manner in which any desired length could be cut. No longer did miners have to use clumsy, dangerously short, unreliable chains of goose quills and lengths of powder twisted in paper; now enough fuse could be cut in preparation for a blast to allow for a safe exit.

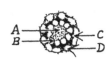

Figure 2. The illustration shows an enlarged cross-section of safety fuse. The core, labeled A, is the powder train, materials B and D are counter-woven layers of fiber, and matrix C is a sealant such as varnish. (Source: E.I. DuPont de Nemours & Co., 1932 p 46.)

Safety fuse proved a godsend from a humane standpoint because it reduced accidents, but also wizened Cornish miners used it to revolutionize mining techniques. They discovered that by trimming fuses into staggered lengths, they could fire an entire group of holes in a set order, rather than firing one to two holes at a time. Shooting groups of holes brought down more rock than miners had seen before, speeding production, lowering costs of mining, and improving safety.

The Industrial Revolution's heavy demand for coal, iron, and other mineral resources spurred the application of blasting on an unprecedented scale, necessitating an explosive which could be manufactured by cheap and simple processes in large quantities. *Black powder*, based on the easy to make, old gunpowder formula more than filled this bill. At this time gunpowder saw several refinements making it better suited for firearms and less so for blasting. Further, gunpowder became prohibitively expensive for the high-volume consumption typical of mining. This was the beginning of the distinction between powder for guns from that which was best for blasting.

By the 1840s the grade of powder manufactured and marketed specifically for blasting became known as *mining powder*. Early mining powder was frequently inferior or reject gunpowder and it was not uncommon for small powder mills to turn out batches of gunpowder which failed quality tests, only to be packaged instead as mining powder.

Saltpeter, chemically known as potassium nitrate, was an oxygen carrier indispensable to blasting powder. With it, blasting powder burned in a self-contained reaction; no outside source of oxygen was necessary. Prior to the mid-nineteenth century, the only sources of usable saltpeter were in the deserts of India, under control of the British. Under British monopoly, the price of saltpeter rose toward the 1850s, and furthermore, shipments were unreliable in quantity, quality, and promptness. These problems caused inventive minds to search for a substitute. It was Lammot DuPont of the famed DuPonts who worked out a solution and patented it in 1857, marking the next significant development in blasting powder[8]. Available in Chile were vast deposits of sodium nitrate saltpeter; however sodium nitrate's tendency to absorb moisture rendered the compound unsuitable[9]. The success to Lammot's patented process lay in refining and specially drying the Chilean saltpeter, and glazing the finished blasting powder with a protective coating of graphite which shielded it from moisture. Lammot named his new powder *B Blasting Powder*, and India saltpeter powder came to be known as *A Blasting Powder*.

Both powders had their advantages and disadvantages. *A Powder* stored longer in humid environments, and it was much better for blasting in moist conditions, such as those found in most mines east of the Rocky Mountains.

In addition, the potassium nitrate saltpeter participated vigorously in the powder's combustion, which resulted in a better blast. *B Powder's* main disadvantage was its vulnerability to moist air in wet mines; after only two to four hours' exposure it could absorb enough moisture to become useless. Further, damp powder was dangerous because of its tendency to produce poisonous gases during combustion. However, because B Powder cost less it was far more popular than A Powder, despite its potentially grave shortcomings. Miners' choice of less expensive B Powder over the superior A Powder exemplified their preference for saving money at the expense of performance.

After Lammot DuPont's breakthrough in the late 1850s, blasting powders changed little during subsequent decades. The last major development in blasting powder was in the form of the endproduct. Following the introduction of *permissible* (government permitted) dynamites to coal mines around 1910, blasting powder began to lose popularity. To make their powders more competitive, manufacturers compressed them into pellets, four of which were wrapped together to form a cartridge like that of dynamite. *Pellet Powder*, the product's name, was introduced to the United States probably by E.I DuPont de Nemours & Company around 1925. By the 1930s it became fairly popular and saw use mainly in coal mines and surface workings where the rock was relatively soft. All of the major powder companies such as DuPont, the Hercules Powder Company, the Atlas Powder Company, and the Austin Powder Company offered their own versions.

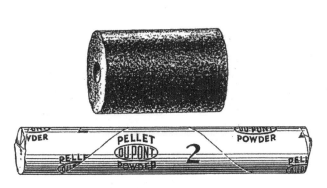

Figure 3. In the mid 1920s E.I. DuPont de Nemours & Company introduced pellet powder *to the United States as an attempt to improve the marketability of blasting powder. Explosives companies compressed powder into pellets such as the unit illustrated, and wrapped four in a waxpaper cartridge. Pellet powder came in three grades for blasting materials of different densities. (Source: E.I. DuPont de Nemours & Co., 1932, p 11.)*

By the 1920s the popularity of blasting powder for mining had passed its zenith. It had already been shut out of hardrock mines in the 1890s, where dynamite proved superior, and permissible dynamites began to displace it in coal mines during the 1910s. Although used in great quantities for open pit mining and quarrying, the consumption of blasting powder declined steadily through the 1930s. After World War II, its demand was so low that all but a few manufactures ceased to mill it, except Atlas, Austin, DuPont, Hercules, and the King Powder Company. By the 1950s the sun had set on blasting powder, which had served as a cornerstone of mining and the Industrial Revolution.

Giant Powder and Its Cousins

The introduction of dynamite to North America by the Giant Powder Company in 1868 began a chain of events that forever changed mining. A very expensive blasting agent at first, dynamite did not make a significant impact on mining until the 1870s, and even then its popularity was limited to the West. But by the 1890s dynamite had proven itself so capable that miners across North America used it almost exclusively for moving rock, and in forty years' time dynamite became the single most popular explosive for coal mining. Between the 1870s and 1900s the mining industry's learning curve of how best to use and handle dynamite was particularly steep. Miners and engineers alike experienced many hard lessons.

Miners generically applied the name *dynamite* to nearly all high explosives. The definition of a high explosive, including dynamite, is a substance that explodes, confined or not, without combustion[10]. This behavior differed from blasting powder, which combusted and exploded only when confined. High explosives are said to *detonate*, which is the rapid decomposition of the solid ingredients, their transformation into gases, and a burst of energy during the phase change.

Although the sudden shock of energy was important for blasting rock, the gases did most of the work. On average, an explosion of dynamite produced over 500 times its original volume in gas, while blasting powder produced only two-thirds of that[11]. This factor coupled with the explosion's shattering nature made dynamite highly efficient for blasting hard rock because it first shattered and then moved the material.

The gases produced by complete detonation of dynamite under ideal conditions were primarily carbon dioxide, water vapor, and nitrogen, with traces of oxygen. When most high explosives were damp or degraded due to age, the above gases changed in percentage and a variety of poisonous and asphyxiate gases were also produced, including toxic nitrous oxide, nitrous monoxide, carbon monoxide, hydrogen, and methane[12]. As conditions in mining and the explosives manufacturing environment were never ideal, in reality the gases produced by the explosion of good, fresh dynamite were a combination of mainly the former group with traces of the latter undesirables.

Straight dynamite, a putty-like mass of nitroglycerine in a combustible absorbent wrapped in a waxed paper cartridge, was the most popular form of high explosive used between the 1870s and 1910s. Miners nicknamed dynamite "Giant Powder," which was the Giant Powder Company's brand name. Straight dynamite certainly was not the first high explosive nor was it the best, but it was the most popular for years.

Dynamite makers produced products based on patented formulas in which the type and chemistry of the absorbent base varied, and the exact quantity of nitroglycerine differed. Yet, manufacturers had to relate the performances of their explosives to a common standard understood by the mining industry. They used straight dynamite as the yardstick to which they compared the performance of other types of high explosive formulas. The stated percentage strength of all high explosive products meant that they acted like their straight dynamite counterpart, not that they contained that quantity of nitroglycerine. In other words, even though a 40% ammonium nitrate-based explosive had much less than 40% nitroglycerine in it, it gave the same strength explosion as straight dynamite with 40% nitroglycerine.

Initiation of an explosion in all high explosives required a shock, while blasting powder, in comparison, required only heat or flame. High explosives were made of molecules joined together by relatively weak bonds, and an explosion happened when the bonds were broken in a rapid chain reaction[13]. *Blasting caps* were the most common means of breaking those bonds. Known throughout the mining industry as *conventional caps, standard caps, detonators,* and *primers,* blasting caps consisted of a copper casing slightly smaller in diameter than a quarter of an inch, and between one and two inches long. The casings were manufactured with drawn construction, resulting in a seamless copper head, but the bottom was open so manufacturers could pack in a mercury fulminate charge, and to allow the cap to slip over the end of safety fuse. By the 1940s cap manufacturers switched to aluminum casings.

The caps sold by mine supply houses in North America during the 1870s were small and barely adequate for detonating straight dynamite. Caps of this vintage came mostly from Europe, particularly Sellier & Bellot located near Prague, in strengths labeled *No. 1* and *No. 2*[14]. By the 1880s American manufacturers, such as the E.C. Meacham Arms Company, the California Cap Company, and the Metallic Cap Manufacturing Company, were selling strengths labeled *No. 3, No. 3 Extra,* and *No.4,* which were strong enough to detonate straight dynamite[15]. By the 1890s cap companies had introduced strengths of *No. 5* and *No. 6* in response to a demand for stronger caps needed to detonate new dynamite formulas. By the mid-1900s cap manufacturers introduced *No. 7* and *No. 8* strengths to their product lines in association with a wave of unconventional high explosives manufactured for the minerals industry. By the late 1910s cap manufacturers dropped all strengths less than

No. 6, and they continued to make No. 7 and No. 8 caps, which remained fairly popular for detonating gelatin, nitrostarch, and ammonium nitrate dynamites. Blasting caps came packed in tins of ten, twenty-five, fifty, and one-hundred, the latter by far being the most common quantity consumed by miners, who used caps by the crateload.

Table 1 Blasting Cap Types, Sizes, and Periods of Relative Popularity[16]

Cap Type	Trade Name	Length in Inches	Compatible Explosives	Relative Popularity
No. 1		5/8	Nitroglycerine, Straight Dynamite	1870 - 1885
No. 2	Double	3/4	Nitroglycerine, Straight Dynamite	1870 - 1885
No. 3	Triple, Triplex, Treble, "XXX"	1	Straight Dynamite, Dynamite Extra, Railroad Powder	1880 - 1915
No. 3 Extra	Extra		Straight Dynamite, Dynamite Extra, Railroad Powder	1880 - 1900
No. 4	Quadruple, Quadruplex, "XXXX"	1 1/8	Straight Dynamite, Extra Formulas, Railroad Powder, Gelatins	1880 - 1915
No. 5	Quintuple, Quintuplex, "XXXXX"	1 1/4	Straight Dynamite, Extra Formulas, Railroad Powder, Gelatins, Low Freezing Formulas	1890 - 1915
No. 6		1 1/2	Straight Dynamite, Extra and Low Freezing Formulas, Railroad Powder, Gelatins, Ammonium Nitrate, Nitrostarch	1900 - present
No. 7		1 5/8	All high explosives	1915 - 1960
No. 8		1 7/8 - 2 1/2	All high explosives	1915 - present

The development of dynamite seemingly happened through a series of isolated events and participants. The first step in the explosive's evolution occurred when a professor of chemistry at the University of Torino, Ascanio Sobrero, discovered nitroglycerine while conducting experiments in 1846[17]. Sobrero found that when he added concentrated nitric acid to glycerin, a very violent reaction ensued and red fumes evolved. He also tried the same with concentrated sulfuric acid and got a similar result. After some trial and error, Sobrero found that by cooling a solution of the two acids and adding glycerin slowly, it dissolved. After a period of time, the solution appeared to be saturated and Sobrero slowly added cool water which caused an oily, viscous liquid to precipitate out. This was nitroglycerine.

Colorless to amber in hue, and viscous, nitroglycerine was based on thick, dense, fatty glycerin, the name being a derivative of the Greek word "glyceros," meaning sweet. Nitroglycerine was heavier than water, it resisted moisture, and it froze at approximately 39° F. These characteristics varied slightly with purity. The explosive liquid degraded and decomposed at temperatures above 120° F, and at 275° F it destabilized, vigorously boiled, and emitted red vapor[18]. If stored for several days in the mid-thirties Fahrenheit, nitroglycerine could form dangerous, unstable crystals. These crystals were unstable in two ways. First, the crystallized nitroglycerine was hyper-sensitive to shock; second, the crystals were in an unstable lattice and readily converted to a highly stable crystal form when the temperature changed[19]. After his experiments, Sobrero discontinued the manufacture of nitroglycerine as he thought it was far too dangerous, but he did secret a sample away in his laboratory.

Young Alfred Nobel first realized nitroglycerine's commercial potential as a blasting agent. He became very familiar with it while he assisted his father, Emmanuel, with explosives research, after four years studying mechanical engineering in the United States. After his father was forced to return to Sweden, Alfred traveled to Paris to round up the capital necessary to establish a small laboratory and manufacturing facility. There, most capitalists laughed at Nobel's enthusiasm for and insistence on the potential applications of nitroglycerine. Despite the almost overwhelming negative response, Nobel managed to scrape up just enough investors to finance the building of a tiny facility at Heleneburg in Sweden. There Emmanuel and Alfred Nobel conducted further research and attempted to make a usable product out of their high explosive. In the early 1860s Emmanuel retired and left the business to Alfred.

Alfred pursued a safe and reliable means of detonating nitroglycerine so it could be used as a blasting product. He knew it was sensitive to shock and intrusions of red-hot metal. He tried detonating it with safety fuse but that only set it on fire. Then one day in 1862 or 1863, while considering the mechanics of the percussion cap in ammunition, Nobel hit upon a bright idea. Why not apply the same concept to detonate nitroglycerine? The result developed by Nobel was a small wooden vial filled with blasting powder and fuse set inside of a vessel of liquid nitroglycerine[20]. The shock of the exploding blasting powder was more than enough force to detonate the nitroglycerine. The idea of the blasting cap was born. In 1863 Nobel substituted mercury fulminate for blasting powder in his cap, and in 1864 he patented a cap of mercury fulminate encapsulated in a lead tube. The final form, patented in 1867, was a mercury fulminate cap in a copper tube, which remained relatively unchanged as the blasting cap for many decades[21]. Although similar in materials and construction, Nobel's cap was weaker than the No. 3 cap miners used for most of the rest of the nineteenth century.

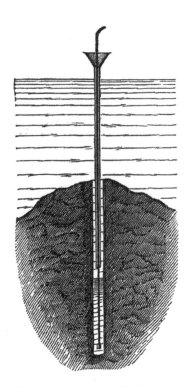

Nobel had a fantastic product and a means of detonating it. All he needed was to make it practicable. As a result, Nobel developed several ways of loading liquid nitroglycerine into drill-holes. For down-holes, the nitroglycerine need simply be poured in through a long funnel, and the detonator floated on top. For horizontal and up-holes nitroglycerine had to be mixed with blasting powder which acted as an absorbent and the detonator inserted into it, or nitroglycerine was contained in tin canisters stopped with a wood plug containing the detonator[22]. By 1864, Nobel had in effect created and patented practical means of manufacturing, loading, and detonating nitroglycerine, the first high explosive product to be used for blasting purposes in mines. These developments made Nobel's high explosive

Figure 4. During the late 1850s and early 1860s Alfred Nobel had developed the use of a long funnel for loading liquid nitroglycerine into drill-holes for blasting. The long funnel facilitated a gentle flow, which was highly desirable when handling nitroglycerine. (Courtesy Hagley Museum and Library; Source: The Nitroglycerine Co. of New York.)

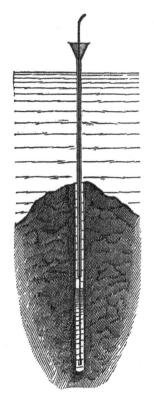

Figure 5. Once the nitroglycerine had been poured into the drill-hole, the blaster inserted a plug containing the detonator over the charge. (Courtesy Hagley Museum and Library; Source: The Nitroglycerine Co. of New York.)

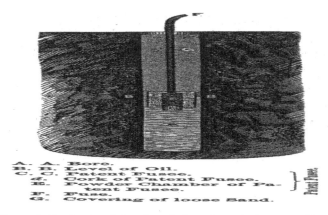

A. A. Bore.
B. B. Level of Oil.
C. C. Patent Fusee.
d. Cork of Patent Fusee.
E. Powder Chamber of Pa-
tent Fusee.
F. Fuse.
G. Covering of loose Sand.

Figure 6. For horizontal and up-holes miners had to pour liquid nitroglycerine into tin canisters with the detonator encapsulated in the stopper. The delicate process of loading the canisters into drill-holes and tamping them in place by the light of an oil lamp or a candle must have been a considerable source of stress and worry for miners. (Courtesy Hagley Museum and Library; Source: The Nitroglycerine Co. of New York.)

product marketable, which was distributed in Europe under the trade name "Nobel's Blasting Oil."

July 15, 1865, was a warm day in New York City. Alfred Nobel and first assistant Colonel Otto Burstenbinder were on hand for a field demonstration of the power of *Nobel's Patent Blasting Oil* to a select group of American capitalists and entrepreneurs[23]. Colonel Burstenbinder poured the liquid nitroglycerine into a drill-hole in bedrock through a long funnel and demonstrated the means of inserting the wooden plug containing the detonator. He lit the fuse and within a few minutes New York City resounded to the first recorded detonation of high explosives in North America. Apparently the demonstration was a great success for almost immediately Alfred Nobel & Co. received orders from mining companies in the United States, especially in California's Mother Lode.

Alfred Nobel & Co. began shipping nitroglycerine to the United States in 1866. At that time…

> Nobel at first claimed that his blasting oil was a 'safe' explosive—that it could hardly be made to detonate except when struck a heavy blow or when confined and ignited by a suitable igniter[24].

Equally, the U.S. Blasting Oil Company, which manufactured under Nobel's rights, claimed in 1866 that nitroglycerine was so safe that it could only be detonated under special circumstances, that it could be stored for an indefinite period of time, and that it could safely withstand temperatures below 212° F[25].

But in April of 1866 the steamship *European* was unloading her cargo in Colon, Panama, so that it could be brought across the isthmus. Included in the lot were seventy cases of liquid nitroglycerine, one of which probably was jarred. In one cataclysm, the ship, the dock, sixty people, and nearby structures were atomized by what must have been a very impressive explosion. In the same month Wells, Fargo & Company's masonry express office in San Francisco was gutted by the explosion of a nitroglycerine shipment. The explosion was caused by several workers who noticed that one of the cases was leaking and began to pry it open, detonating the loose nitroglycerine[26]. Later, it came to light that although Nobel's nitroglycerine was made of the best ingredients at the time, they were none the less impure and over time the nitroglycerine destabilized. After these and several other disasters, freight companies refused to ship nitroglycerine, which excluded Alfred Nobel & Company from the North American market.

Alfred Nobel promptly journeyed to the United States in the Spring of 1866 to remedy the situation. By summer Nobel had signed a contract with industrial capitalists Israel Hall, James Deveau, Timothy Church, and Edgar Wait to manufacture up to 1,000 pounds of *blasting oil* per day under a U.S. patent. The new organization incorporated themselves in 1866 in New York as the United States Blasting Oil Company, the first high explosives maker in North America.

Immediately after obtaining a patent and establishing the United States Blasting Oil Company, Nobel hurriedly returned to Sweden to develop better packaging for it. Up to that time, Nobel had been shipping nitroglycerine-filled soldered tin cans and bottles in wooden boxes with sawdust padding. After opening cases returned because of suspicious stains, he discovered nitroglycerine leaked out of the cans and into the sawdust, causing the wood to oxidize badly. Further experiments showed that sawdust saturated with nitroglycerine had the potential to smolder and spontaneously catch fire. This, no doubt, was the cause of at least some accidents in the past.

To alleviate this problem, Nobel searched for an inert substitute which he found in the form of a diatomaceous earth known as *kieselguhr* near his main plant at Krummel, Germany. The fact that the kieselguhr worked perfectly as a safe packing material for the cans was brought to his attention through a letter he received. It seemed that a customer opened a perfectly good, clean case of blasting oil to discover that at least one can leaked, and that the kieselguhr absorbed all of the loose nitroglycerine. The case was returned to Nobel, and after close inspection he surmised the customer was right in that the kieselguhr totally absorbed the nitroglycerine[27]. At once Nobel realized the significance of kieselguhr's absorbent qualities and after investigations he found that this new explosive mass was easily cartridged and packaged in waxed paper, which meant no more messy liquid nitroglycerine for blasting. In 1867 Nobel took out the first patent for this new high explosive, which he named *dynamite*, taken from the Greek work for power, *dynamis*.

For Horizontal Bores.

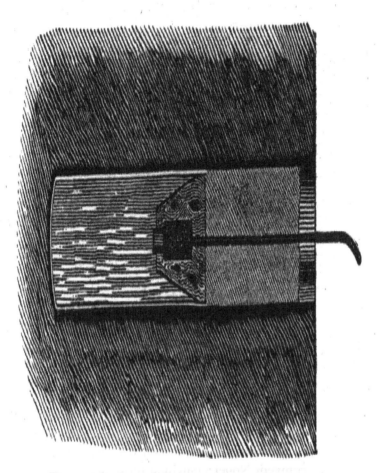

1st. For such bores, cartridges must be used. The cartridges are filled with the oil nearly to the top.

Figure 7. The Nitroglycerine Company, successor to the United States Blasting Oil Company, used an appropriately ominous trademark to identify its products. This advertisement dates to the early 1870s when liquid nitroglycerine was used in a few hardrock mines. (Courtesy Hagley Museum and Library; Source: The Nitroglycerine Co. of New York.*)*

The invention of dynamite, as earthshaking as it was, happened quite by accident.

Although dynamite was a vast improvement over liquid nitroglycerine, it had its drawbacks. Through trial it was found that kieselguhr had a loose purchase on the nitroglycerine, which occupied 75% of the mass by volume[28]. Water easily displaced the nitroglycerine, and if allowed to stand long enough in storage, the nitroglycerine gravitated to the bottom of the cartridge. Also, because kieselguhr was inert, it did not contribute to the explosion, and in fact it tended to absorb some of the energy. Despite its drawbacks, kieselguhr was a keystone. The first dynamite sold in North America included the inert absorbent, and it served as a springboard for further development.

Figure 8. In February 1869 the San Francisco firm Bandmann, Nielsen & Company, the Giant Powder Company's agents and financiers, submitted the first advertisements published in North America for dynamite. The ads were run in several mining journals, including the Mining and Scientific Press, *in hopes of creating interest. Dynamite was far from an immediate sensation in the mining industry. (Source:* Mining and Scientific Press, *1869.)*

During the 1860s, Nobel's European-based explosives company remained cut off from North America. In response, Nobel traveled to San Francisco to patent his dynamite and establish a subsidiary manufacturing facility. In 1867 Nobel enlisted Western capitalists, businessmen, and chemists, and they formed the Giant Powder Company, which held rights to Nobel's invention in the United States[29]. After "Giant Powder" began experiencing popularity in the West's mines, other American entrepreneurs examined Nobel's dynamite, and they attempted to make their own high explosive product while trying to circumvent Nobel's patented formula.

The problems and limitations of using the inert kieselguhr absorbent were recognized early by independent chemists and inventors, and they attempted numerous improvements. Carl Dittmar patented the first dynamite with an *active-base* in 1867 in Germany, and his formula served as the next big development in high explosives[30]. Some sources erroneously claim the inventor of dynamite with an active-base was Nobel, but his was a British patent granted in 1869, which followed Dittmar's by two years. Dittmar's active-base was a wood meal which absorbed nitroglycerine, and it was a great improvement over inert kieselguhr because it held onto the liquid explosive better and it actually participated in the explosion. This was a

Figure 9. Around 1874 the Giant Powder Company began shipping dynamite in wood boxes containing 50 pounds, and stenciled with labels as illustrated. Giant's label format changed little until the late 1880s. At approximately the same time the California Powder Works began shipping its dynamite in boxes conforming to the second illustration. The California Powder Works used the same label style into the early 1880s. (Source: Author.)

milestone for mining, because *straight dynamite*, the most popular dynamite from the 1870s until the 1910s had arrived. Chemists and inventors formulated and patented a number of variations, but Dittmar's concept of an active base applied to all. In fact, Dittmar's wood meal was used by some manufacturers as late as the 1900s. Within a few years of Dittmar's invention, American chemists began developing their own dynamite with active bases. The general trend was to use compounds which significantly contributed to the explosion, most notable of which was saltpeter[31]. During the 1870s Western mining companies began using the new explosive, although the switchover from blasting powder lasted into the early 1890s in some regions[32].

Dynamite and blasting powder proved different in nearly every respect from one another except their fundamental functions as blasting agents. Dynamite required different storage, handling, and preparation techniques, and its behavior was different. Its manufacturers passed on some of the basic information regarding dynamite to mining companies, but the information was rarely conveyed to miners. This lack of communication, and the fact that explosives makers knew little about the long-term behavior of dynamite ensured that miners and mining engineers learned about the blasting product primarily by trial and error, close calls, and death. The generation of miners that experienced the switch from blasting powder to dynamite during their active careers had the greatest difficulty. They were intimate with blasting powder, having used it for some time, and the different nature of dynamite required a change in habits.

An anecdote from Michigan's iron mines serves as an example. Titus Hibbert, superintendent of the day shift at the Jackson Iron Mine in Michigan, was instructed by the mine boss to test a batch of dynamite to see if better results could be had than with the blasting powder they were using. Hibbert stuffed his pockets with ten cartridges and rode the ore bucket down the shaft where he met his subordinate level foreman in an exploratory drift. The foreman, who also had never used dynamite, accepted the sticks and said with great trepidation, "Who ever heard of stuff like that beating good old black powder?" The level foreman primed it as instructed, loaded it into the face, lit the fuse and retreated into a crosscut where the drilling crew was waiting. After the round shot, Hibbert and the level foreman returned to investigate. The face did not have the clean-cut look typical of a successful blast; it was a concave mess. Upon inquiry, the level foreman admitted to loading all ten cartridges, enough to shoot the face, into one center hole because he thought "dynamite couldn't possibly equal blasting powder"[33].

There are two points to this anecdote. First, the level foreman, an experienced miner, clearly displayed his ignorance of dynamite, so typical of miners in transition from blasting powder. Dynamite proved itself time and again to be better than blasting powder in hard rock, but due to miners' prejudices and reluctance to change, this point was not always received. Many miners did not trust dynamite - they thought it was inferior to blasting powder, or at best treated it like blasting powder, which had grave potential. Enticing miners to switch to dynamite was quite a battle. In fact, some miners mixed dynamite cartridges with blasting powder in the same drill-holes and set off the mass with an old-fashioned squib, which was all wrong[34]. The other point raised by the above anecdote is that many miners treated dynamite with the same safety regards as blasting powder. Because they did not understand it and how it differed from blasting powder, they did not understand its unique hazards.

One of dynamite's most notorious temperaments was its seemingly unpredictable state of stability. Early dynamite was enormously less sensitive to shock than pure nitroglycerine, but it was extremely sensitive none the less. Part of the reason for this was that the nitroglycerine began to settle to the bottom of the paper cartridge when in storage. Two miners were preparing a batch of old dynamite cartridges on the 400 level of Cripple Creek's Gold Sovereign Mine. Either they were piercing the cartridges to insert blasting caps, or they were slitting the waxed paper wrappers so the cartridges could expand when tamped into drill-holes. In either case, the dynamite was so sensitive that one of the cartridges went off, in turn detonating all forty pounds the miners were working with[35]. For this sort of behavior dynamite gained a bad reputation. Take dynamite's sensitivity and its combustible nature, add to that miners' lack of training and experience and in some cases a devil-may-care attitude, and the foundation was laid for disaster.

Straight dynamite, the only kind of dynamite widely available from the early 1870s until the 1910s, froze and became unusable between 40° F and 50° F. The ambient temperature in many regions in North America became that cold even in summer, especially the Northern and Mountain States, where hardrock mining thrived. Attempting to detonate frozen dynamite amounted to high offense in underground mining, because only a portion of the charge exploded. Partial detonation resulted in two problems that threatened miners. First, poisonous gases were given off, and second, some unlucky miner had to search for the remaining live cartridges. Thawing dynamite was absolutely necessary, and when done improperly, it was quite dangerous because of the risk of setting the dynamite off. The following event, which occurred high in California's Sierra Nevada Mountains, neatly illustrates what could happen when untrained miners attempted to thaw dynamite to a usable state:

> A frightful accident occurred at the Great Sierra Mine, Tioga District, about 11 a.m. on Thursday last (1881), by which three men were seriously, one probably fatally injured. It is the old, old story of thawing frozen nitroglycerine powder on a stove. A short distance east of the shaft and not far from the blacksmith anvil stood a large box stove, which was usually kept glowing with heat. In, on, or about this stove someone placed six sticks of frozen Excelsior powder…the powder exploded with a terrific crash, tearing away the outer wall of the blacksmith shop and knocking the men senseless. James Kickham was severely and probably fatally injured, receiving a ghastly wound on the head, several lacerations of the arm and body, and being literally torn to pieces about the pelvic region[36].

It has been said that fully one-quarter of all accidents involving dynamite occurred during thawing[37]. In fact, thawing accidents were so frequent that municipalities and states passed of laws regulating thawing as early as the late 1880s.

As can be expected, miners and engineers differed on the best means of thawing frozen dynamite. Miners employed methods that incorporated convenience, simplicity, and required little money, and they tended to be unsafe. Mining and explosives engineers, on the other hand, endorsed slow, capital-intensive but safe practices. Specifically, depending on the mine's consumption of dynamite, they recommended either thawing kettles, thawing boxes, specially constructed steam-heated magazines, or thaw-houses. *Thawing kettles* had watertight inner chambers made of zinc-coated steel with a capacity of up to twenty-five pounds of dynamite. Some had tubes for individual cartridges and others consisted of single chamber, and both were surrounded by a hot water reservoir. *Thawing boxes* featured watertight chambers like kettles, but they were capable of holding up to fifty pounds of dynamite. Some boxes were steam-heated through pipe couplings. Well

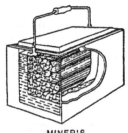

BRADFORD
THAWING KETTLE

CATASAUQUA
THAWING KETTLE

MINER'S
THAWING KETTLE

Figure 10. The illustration shows the three most common dynamite thawing kettles. In most cases, metal-working factories wholesaled kettles to explosives makers and mine supply houses, which in turn sold them to miners and mining companies. The Bradford *and* Miner's *styles each held approximately twenty-five pounds of dynamite and had spouts for pouring hot water into the jacket. All the types featured handles so miners could carry the dynamite to their workings while it thawed. (Source: Peele, 1918, p 161.)*

capitalized mines that consumed large quantities of dynamite erected special steam-heated buildings where miners deposited entire unopened boxes for thawing. Most explosives engineers insisted that these were the only safe thawing methods and that nothing else was acceptable.

Thawing devices cost money, which most self-financed and contract mines

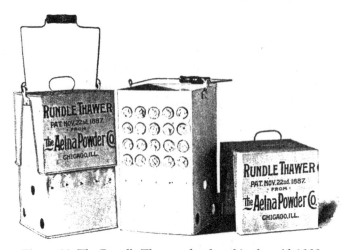

Figure 11. The Rundle Thawer, developed in the mid-1880s and manufactured into the 1910s, was one of the few factory-made thawing kettles to feature individual tubes for holding cartridges. Hot water, poured in through the top, surrounded the tubes, and a candle could have been placed under the thawer's bottom to keep the water warm. Shop workers employed by mining companies used the Rundle as a model for making imitations. (Courtesy of Hagley Museum and Library; Source: Aetna Powder Co., 1900.)

and miners did not want to spend. Still, their dynamite had to be thawed, and in response they sought alternative methods which were at least highly

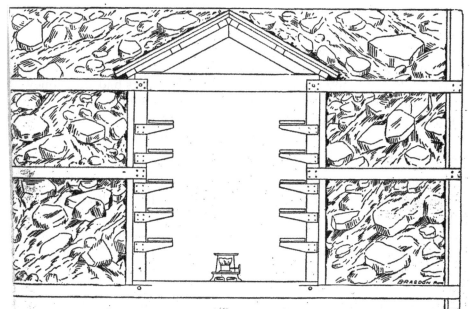

Figure 12. A cut-away view of a combination earthen-covered dynamite thaw house and magazine. The soil backfill encasing the structure served as insulation, and the central kerosene stove provided heat. The magazine keeper placed unopened boxes of dynamite on the shelves where they stayed warm. Facilities such as this cost mining companies time, materials, and space, which smaller or low-budget operations typically could not or would not afford. (Courtesy Hagley Museum and Library; Source: Kirk, 1891, p 11.)

questionable. According to one engineer: "The most dangerous means of thawing cartridges are ingeniously devised by ignorant laborers; baking, boiling, and toasting being favorite methods"[38]. Miners attempted to thaw dynamite in hot sand, placed before fires, warmed on cook stoves, heated by hand over miners' candlesticks and oil wick lamps, and even in their bootlegs! The blacksmith shop was a favorite place of thawing as there were a number of warm places such as the forge, wood stoves, and the quenching tank, as a blacksmith at a mine in Bergen, New Jersey, found when he plunged red-hot drill-steels into his tank, detonating free nitroglycerine that settled out there[39]. It is no wonder that the blacksmith shop was frequently the site of thawing accidents. Another favorite means of thawing was to dunk frozen dynamite into boiling water. In one case a quarry crew was steaming their dynamite over boiling water. They had a low fire and plenty of water so the dynamite would not overheat, and they heeded the advice of engineers not to outright boil it. After a short time they got quite a surprise when their entire thawing apparatus was blown to atoms. What they did not know was that thawing, especially in hot water and steam, caused nitroglycerine in straight dynamite to spontaneously exude, so it settled to the bottom of their water pot, overheated, and exploded[40].

These methods of thawing prompted one mining expert to proclaim, "If the dynamite freezes don't thaw it in the cook stove oven or in a frying pan over the blacksmith's forge, and don't plunge it into boiling water. Better to have a small kettle which you can buy from your dynamite dealer"[41]. But in many mining districts even if miners had the money to spend, thawing kettles were often difficult to obtain. Some mining engineers with a sympathetic slant toward the miner felt that mining companies should provide their miners with thawed dynamite, from a humanitarian and production stand-point[42].

True salvation for miners came in the form of low-freezing dynamite formulas which underwent commercial development in the 1890s. Alfred Nobel and other chemists quickly acknowledged the problems caused by the high freezing point of nitroglycerine, and they attempted to solve them as early as the 1870s. But the technology of blasting caps, chemistry, and the manufacture of dynamite, as well as marketing, were not sophisticated enough to facilitate success. F. Broberg patented what may be the first successfully marketed commercial low-freezing dynamite in United States in 1895[43]. It was a naphthalene-based formula with which his company, the Nitro Powder Company, experienced limited success.

Shortly after 1900 S.H. Flemming of the E.I. DuPont de Nemours Powder Company patented another practical low-freezing formula which utilized a blending of nitroglycerine and TNT, bringing its freezing point down to about 30° F. DuPont's formula, sold under the *Red Cross* brand name beginning in 1906, did very well because of a fusion of three factors. First, after decades of lost limbs and lives the mining and quarrying industries were quite ready for solutions to the thawing problem. Second, like preceding formulas, DuPont's low-freezing dynamites did not detonate as easily as conventional straight dynamite, but by 1910 No. 6, No. 7, and No. 8 blasting caps were widely available, and using them ensured detonation. Last, by the time DuPont developed its low-freezing dynamite the company operated twelve plants across the nation, and such a manufacturing and distribution capacity brought the product within reach of many mining operations.

Explosives consumers latched onto the concept of low-freezing dynamite, and the blasting agent began to arrive by the freight-car load in cold-climate mining districts beginning in the 1910s. During that time all major explosives manufacturers, including the Hercules Powder Company, the Atlas Powder Company, DuPont, the Aetna Explosives Company, and the Keystone National Powder Company, offered low-freezing variations of DuPont's formula. These low-freezing products were readily accepted at many mines, but because of their higher cost, and inertia against change, miners continued to use old-fashioned straight dynamite, albeit in decreasing amounts. By the 1920s dynamite thawers gathered dust as miners ubiquitously used low-freezing formulas in cold climates. The age of thawing dynamite over candle flames, in shirt pockets, and on stoves had mercifully come to an end.

Another of dynamite's quirks that miners learned by trial was its unpredictability once on fire—a phenomenon miners were more than capable of achieving. A cartridge of dynamite consisted of an absorbent akin to blasting powder wrapped in waxed paper, which in sum was highly flammable. Given that many miners exposed dynamite cartridges to open flame in their attempts to thaw the explosives, the fact that they set cartridges alight comes as no surprise.

Usually burning dynamite proved catastrophic, but not always. Shortly after 1910 lessee miners T.J. Dalzell, Billy Beard, and two others were riding an ore bucket down a winze in Cripple Creek's rich Vindicator Mine. Beard was smoking a cigar and two other miners were enjoying their pipes during the decent. Miners' tools lay at their feet in the bucket, including a burlap bag containing eighteen dynamite cartridges, blasting caps, and fuse. Beard disembarked at Level 3, placed the bag of dynamite in a partially full box already at the level, and walked down the drift to resume his work. He was running a rockdrill when he noticed a bright light where there was previously none. Beard instantly knew what the light was—his box of dynamite was on fire—and it lay between him and safety. He dropped what he was doing and began a mad dash down the drift, skirting the flaming box, and as Beard was headed away, in the dim light he tripped and fell. As he hit the drift floor the box of dynamite exploded and the blast went over him, wrecking a portion of the shaft station but leaving Beard unscathed. What had happened during their decent was that an ember or smoldering match had fallen from one of the smoking miners into Beard's burlap sack. It ignited the flammable dynamite after Beard had placed the bag in the partially full box[44].

In another instance, during construction of the New York City subway in 1901, a worker on the night shift left a burning candle in a tool shed for light. Unfortunately, the candle fell over after the worker left and started a fire. As soon as the workers saw the flames they scattered in all directions, fleeing for their lives, because the tool shed doubled as their dynamite magazine. The dynamite, well-thawed, caught fire and detonated, blowing the shed and environs into splinters[45].

However, the behavior of burning dynamite was not always predictable. One morning two miners in Montana began to thaw some dynamite to be used that coming day. They laid it:

> against their open oven door so the dynamite could thaw while the men ate breakfast in their cabin on a frigid winter morning. Suddenly the men realized the dynamite had caught fire and 'was blazing and

sputtering at a furious rate.' Fearing that the two full boxes nearby would be detonated, they raced out the door and sat on the opposite slope, expecting the cabin to be blasted to pieces. Instead it burned to the ground...[46].

Why did the two boxes of dynamite not explode? The answer will never be known for certain, but in many cases small batches of thawed dynamite burned quite well without exploding[47]. But as demonstrated by Billy Beard, the miners in the cabin, and the subway workers, burning dynamite was wisely feared by all.

Although straight dynamite was the most popular high explosive for some time, it had shortcomings. For example, it did not perform well when completely submerged in water because water displaced the nitroglycerine and spoiled the base. In addition, straight dynamite produced high volumes of noxious and poisonous gases, and the explosive was not always strong enough to shatter very hard rock. In terms of these characteristics, other high explosives proved better, among which were gelatin and ammonium nitrate.

Gelatin, known before 1915 as *gelatine*, was a nitroglycerine-based dynamite invented by Alfred Nobel in 1873. Nobel was at work in his laboratory developing active bases for dynamite, and after a little experimentation he found that guncotton, a form of nitrocellulose, readily dissolved in nitroglycerine, and when he allowed the solution to stand for a number of hours, it coagulated into a gel. As a blasting agent, gelatin was well-suited for mining. The explosive manifested as an elastic putty with excellent water resistance, it tended to produce the least harmful gases of most commercial high explosives, and its velocity of detonation was quick.

When Nobel developed his new gel into a commercial explosive, he found that the substance lent itself well to the creation of both *blasting gelatin* and *gelatine dynamite*. Nobel patented the formulas in Europe in 1875, and in the United States in 1876[48]. Nobel found that the performance, usability, and water

 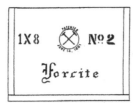 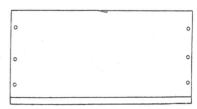

Figure 13. The American Forcite Powder Manufacturing Company was the first explosive maker in North America to make gelatine, later known as gelatin, dynamite. This fifty pound box was American Forcite's standard shipping container between 1883 and the 1890s. Due to its higher cost and its unconventionality, few miners used gelatin until the 1910s. (Source: Author.)

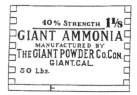 

Figure 14. Giant Ammonia *was one of the few ammonium nitrate dynamites to enjoy success prior to the 1920s. The Giant Powder Company, Consolidated packed its ammonia dynamite in boxes identical to the one illustrated from approximately 1907 to 1915. (Source: Author.)*

Figure 15. Explosives manufacturers usually sold *railroad powder in bulk form, which they packed in waxpaper bags like the one illustrated. Bags came in twelve-and-one-half. or six-and-one-quarter pound sizes, which were packed in a fifty pound box. The Repauno Chemical Company, one of the United States' leading explosives makers, produced the illustrated bag during the 1890s. (Courtesy Hagley Museum and Library; Source: Repauno Chemical Co., 1895.)*

resistance of gelatine was much better than regular dynamite for blasting hard rock and for work in wet conditions. Although the Belgian Force Works manufactured gelatine dynamite almost immediately in Europe, gelatin was not made in the United States until the American Forcite Powder Manufacturing Company started up its nitroglycerine nitrators in 1883[49].

Despite gelatine's early introduction, the explosive did not experience popularity in America until the 1910s. In terms of production, gelatin was dangerous to manufacture, miners found the explosive insensitive to the No. 3 blasting caps commonly available at the time, and it was more expensive than straight dynamite. The innovative explosive began to experience mild popularity in the 1890s. By approximately 1910 gelatin's popularity surged, and by 1915 miners favored it over straight dynamite.

While hardrock miners overwhelmingly preferred straight dynamite between the 1870s and 1910s, they also used several other blasting agents which are worthy of mention. During the late 1860s, visionary chemists realized the potential that Nobel's dynamite held. In hopes of competing, they began tinkering with alternatives to nitroglycerine in hopes of producing an equally effective explosive. Chemists Johan Ohlsson and Johan Norbin patented the first of these, an ammonium nitrate formula, in Sweden in 1867 under the name *Ammoniakkrut* (*Ammonia Powder*)[50]. The first formula they produced included 80% ammonium nitrate and 20% charcoal, but the

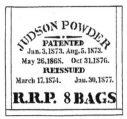  

Figure 16. The Judson Powder Company, which existed between 1881 and 1888, was the first explosive company to make railroad powder, and it set the industry standard for packaging. Judson originally packed four or eight waxpaper bags in 50 pound boxes like the one illustrated (Source: Author.)

ammonium nitrate had a great affinity for moisture, and it was too insensitive to be of practical use. Under advice from their friend Alfred Nobel, they substituted 14% nitroglycerine for some of the charcoal, which sensitized the compound and added water resistance. They packed the compound in zinc boxes and India-rubber bags to guard against ever-present moisture, but it still did not perform well enough. In 1873 Nobel bought the patent rights, and rearranged the formula to include either paraffin, stearic wax, or naphthalene to further slow water absorption. Yet, the explosive still did not perform well enough to compete on the market, so he tucked it away pending further improvements.

Always thinking about possible applications of his products, Nobel hit upon the idea of using a liquid-like gelatin instead of nitroglycerine to solve ammonium nitrate's moisture problem. Nobel filed a Swedish patent for *Extra Dynamite* in 1879, and in so doing he was the first to produce a commercial high explosive based on ammonium nitrate[51]. Extra dynamite, also known as *dynamite extra* and *special dynamite*, was an improvement over straight dynamite. The ammonium nitrate served as an absorbent for nitroglycerine, and it was explosive itself. The combination resulted in a more powerful, longer explosion than straight dynamite, and its byproduct gases were cleaner. However, water easily spoiled the ammonium nitrate compound, and when wet the explosive produced noxious gases. For moist conditions *gelatin extra*, a cousin to dynamite extra, served better.

Although dynamite extra was developed fairly early, it experienced a latency period before it became popular. The Giant Powder Company began manufacturing dynamite extra as early as 1885, and the Atlantic Dynamite Company possibly several years before that. Still, not until 1905 did extra formulas see even mild sales. By 1915 sales of extra formulas blossomed, and, supplementing gelatin dynamite, replaced the old straight dynamite miners had used underground[52]. The reasons for the abrupt and delayed success of extra products are the same as those behind gelatin's success.

In 1885 American chemists Russell Penniman and John Schrader, employed by the DuPont syndicate, became the first to formulate a commercial explosive based on pure ammonium nitrate. Penniman and Schrader developed and patented a process in which ammonium nitrate was dried, granulated, and each particle was coated with a petroleum-based compound such as Vaseline, cosmoline, or crude oil to repel water, and the resulting mass was mixed with a conventional active base such as sodium nitrate[53].

Like other alternative high explosives in the late nineteenth century, ammonium nitrate was at first more costly than straight dynamite, it too required at least a No. 5 blasting cap for detonation, and it had the drawback of being new and unusual. These characteristics rendered the explosive unattractive to all but the most progressive mining operations. The Giant Powder Company, Consolidated was the first explosives manufacturer to produce significant quantities of pure ammonium nitrate dynamite, beginning around 1907. Until that time, explosives manufacturers such the California Powder Works, the Repauno Chemical Company, and the Atlantic Dynamite Company used Penniman and Schrader's ammonium nitrate to make extra dynamite.

For miners, ammonium nitrate was both a burden and a blessing. The explosive was a significant improvement over nitroglycerine dynamites when it came to handling. Instead of becoming increasingly sensitive and hence dangerous in storage, ammonium nitrate grew less so. Nitroglycerine gave the most pulsating headaches upon dermal contact, which happened occasionally when miners primed dynamite cartridges. Not so with ammonium nitrate. The innovative blasting agent also possessed significant drawbacks. The potential to produce poison gases was high, contingent on how dry miners kept it. But when treated properly, ammonium nitrate had some of the cleanest byproduct gases of any explosive. For all of these reasons ammonium nitrate became popular by the 1910s.

Perhaps ammonium nitrate's greatest contribution to the mining industry was as a low-priced, free-flowing bulk explosive for open pit mining. Its explosive qualities made it ideal for moving huge quantities of rock in the large blasts practiced in open pit mines. Ammonium nitrate's physical state was conducive to selling it in bulk form, which open pit mining companies needed for loading into deep drill-holes. Ammonium nitrate's explosion started much quicker than blasting powder, but proceeded a little slower and was more heaving than nitroglycerine dynamites. In the 1920s ammonium nitrate began to compete with blasting powder and railroad powder, the two most popular explosives in open pit mines, and by the 1950s it had shoved blasting powder off to the side in terms of popularity.

Nitrostarch was the only other commercial non-nitroglycerine explosive that experienced widespread popularity among America's miners. Nitrostarch was superior in many ways to straight dynamite and ammonium nitrate. Physically, nitrostarch possessed characteristics similar to ammonium nitrate. In its pure form, nitrostarch detonated with effects similar to liquid nitroglycerine, and because of this, it had to be mixed with an active base to modify the explosion into something usable for hardrock mining. Typically, nitrostarch formulas consisted of 30% nitrostarch, 15% ammonium nitrate, 47% saltpeter, and traces of charcoal and other ingredients.

One of the reasons nitrostarch caught on in the mining industry was that it offered benefits similar to those of ammonium nitrate. When dry, the byproduct gases produced by nitrostarch were much cleaner than straight dynamite, and it was not as susceptible to moisture as ammonium nitrate. Because nitrostarch included no nitroglycerine, and because it decomposed over time, it was safe to handle and store. Many miners grew to appreciate the merits of nitrostarch, and its popularity grew through the twentieth century. Like ammonium nitrate, nitrostarch lent itself well to open pit mining because it too was easily packaged in bulk form and in large-diameter cartridges.

The history of nitrostarch is inseparable from explosives chemists F.B. Holmes, and J.B. Bronstein and their Trojan Powder Company. Holmes and Bronstein began experimentation in a laboratory in Allentown, Pennsylvania, where they sought a safe and efficient process to manufacture the explosive in commercial quantities. In 1903 they succeeded and immediately organized the Allentown Nonfreezing Powder Company. Allentown's profitability and potential was encouraging, so the chemists reorganized it as the Pennsylvania Trojan Powder Company in 1906, and consolidated eastern and western manufacturing plants into the Trojan Powder Company in 1918[54]. Trojan was the only company to successfully produce nitrostarch, and while it had a hard road competing against the likes of the Hercules Powder Company, the Atlas Powder Company, and DuPont, the company experienced success in

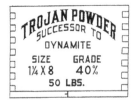

Figure 17. The Trojan Powder Company, the only significant maker of nitrostarch, promoted its products as being superior to nitroglycerine dynamites, as suggested by the statement on the boxes the company used circa 1920s. (Source: Author.)

hardrock and open pit mines through the Great Depression and into modern times.

Alfred Nobel's search for commercial dynamite formulas in the early 1860s yielded a specialized high explosive which found favor decades later. In his experiments, Nobel attempted to use blasting powder as an absorbent in a nitroglycerine dynamite. He shelved this formula because he found blasting powder to be a poor absorbent, and that it significantly dampened the explosion. Years later, in 1871, the California Powder Works made a similar product called *Black Hercules* to compete with straight dynamite made by the Giant Powder Company. The California Powder Works quickly dropped production of this formula because the nitroglycerine-blasting powder compound was much slower-acting than Giant's quick, shattering dynamite and, hence, not as effective in hard rock. It did not meet miners' expectations of dynamite.

Egbert Judson, one of the Giant Powder Company's visionary founding fathers, duly noted the performance of the mixture of a small volume of nitroglycerine absorbed in blasting powder, and its possible applications. In 1876 he patented *Judson Powder*, which consisted of a modified blasting powder thinly coated with nitroglycerine[55]. The difference between Judson's product and earlier nitroglycerine-coated blasting powder formulas lay in the manufacturing process. Judson melted the sulfur, added the other powder ingredients to the melt, cooled it, crushed it, and classified the grains. Melting the ingredients made their surfaces glassy, decreasing porosity and facilitating an even coat of nitroglycerine. In the making of true blasting powder, there was no melting of ingredients. Judson's formula required the grains be coated with 5%, 10%, or 15% nitroglycerine by weight.

Judson Powder's explosion fell somewhere between a slow dynamite and the fastest blasting powder. Because of this performance, it quickly gained favor over both dynamite and blasting powder for work in ground which was an inconsistent mix of rock and compact sediment, and in soft, sedimentary rocks. Such blasting conditions were commonly encountered in open pit mining, coal mining, land development, hydraulic placer mining, and railroad construction. In fact, Judson Powder became so popular for railroad construction that it earned the name *Railroad Powder*. Like ammonium nitrate, the low price, performance, and packaging of railroad powder in bulk form or cartridges made it perfect for open pit mining, and by the 1910s these mines consumed railroad powder in the thousands of tons.

As soon as Judson organized the Judson Powder Company in New Jersey in 1876, other manufacturers imitated his railroad powder and the product proliferated. In short order it found its way underground where miners used it to shoot soft material, especially coal. By the 1900s mining and explosives engineers condemned it because it created more poisonous gases than did straight dynamite. Any time nitroglycerine and blasting powder were combined, as railroad powder did so well, combustion of the powder burned some of the nitroglycerine and the waxed wrapper, resulting in foul gases. Still, because of its low price coal miners continued to use it into the 1920s.

"Safety" Explosives

No one could dispute that working in a mine was dangerous. Hardrock miners had to contend with injury or death from a host of threats, but coal miners had an additional worry—the threat of fire and explosion that could turn an entire mine into an oxygen-deprived hell. Coal mines often contained two highly flammable materials: well-trampled, well-pulverized, fine coal dust, and naturally occurring methane gas, which commonly seeped into mine atmospheres from the surrounding sedimentary geology. Ignition of both of these materials had caused many disastrous mine explosions and fires, and the U.S. Bureau of Mines proved that the heat and flame of explosives, especially blasting powder, were often culprits[56]. Methane was ignitable at temperatures above 650° F while commercial explosives typically detonated at temperatures well above 1,000° F. The U.S. Bureau of Mines studies also concluded that shock waves generated by the detonation of dynamite suspended near-by coal dust in some instances, making it even more vulnerable to ignition.

The hazards of using explosives presented the coal mining industry with a considerable problem: how to maximize production without banning explosives underground and returning to brutal, slow manual labor? The problem was not isolated to North America; the European coal mining industry experienced the same dilemma. However, European explosives and mining industrialists acknowledged and began tackling the problems associated with safe blasting in coal mines before those in America did.

The first attempts at cooling explosions in gaseous and dusty mines were devised in England in the 1880s. Some contrivances were awkward and almost humorous, such as a special sprinkler system which showered working faces in mines with water during blasts, and attempts at tamping saturated moss into blast-holes to absorb some of the raw flame of an explosion. The closest things to real solutions aimed at minimizing open flames created by blasting included loading special water canisters on top of charges, which would rupture during the blast, and an explosive charge suspended by sheetmetal spokes in the center of a water-filled canister[57]. These primitive attempts at suppressing blast flame were rather inconvenient to use and probably not very effective.

The first practical solution was the rational approach of modifying the explosive rather than the blasting environment. It came in the form of *Carbonite*, a *safety explosive* patented in 1887 by a Mr. Bichel in Germany. Carbonite was a breakthrough for the explosives industry because although it was based on straight dynamite, Bichel cooled its detonation temperature by incorporating carbonaceous matter. Shortly after Carbonite was developed, another explosives chemist patented *Monobel* in England.

It was not until the 1890s that American explosives manufacturers began serious experimentations designed to minimize the explosion temperature of their dynamites. One of the first explosives chemists to develop a viable product was General Paul Oliver, who developed

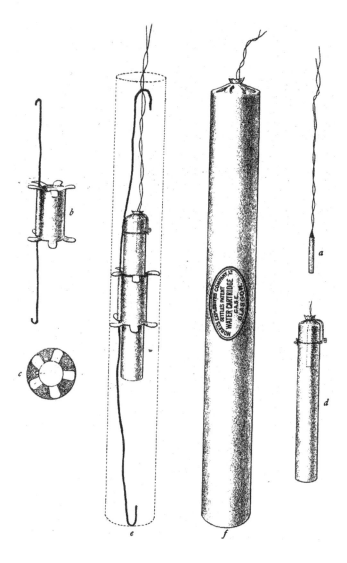

Figure 18. "Settle's Patent Gelatine Water Cartridge," patented in England and possibly manufactured during the 1880s, encased a dynamite cartridge in a water jacket to squelch flame when used in gaseous mines. Curiously, the device specified the use of gelatine dynamite, which was one of the least desirable explosives for blasting coal, probably because gelatin had the greatest resistance to the water in the canister. Note how the cartridge has been primed with an electric blasting cap, at a time when electric blasting technology was nascent. (Courtesy Hagley Museum and Library.)

Oliver's Flameless Powder in the mid 1890s. Oliver's formulas consisted of blending hydrated salts with low percentage straight dynamites. His Luzerne Powder Company, based in Laurel Run, Penn-sylvania, had modest success with the safety explosive, prompting Oliver to introduce two other short-flame explosives with even less nitroglycerine, under the names of *Blackjack* and *Victor Coal Powder*. Surprisingly, despite very successful performances, Oliver's products saw only local popularity.

The years 1902 to 1905 saw a number of safety explosive patents filed by chemists—a sign that both the demand and manufacturing technology for safety explosives were approaching an intersection point. Beginning in 1902 DuPont undertook serious research and development of short-flame explosives and sent its chemical director, Dr. Charles Reese, to Europe to study state-of-the-art formulas and manufacturing processes. Two years later, DuPont established its Eastern Laboratories at the sprawling Repauno Plant in New Jersey, where chemists and engineers began addressing the issues of practical manufacture of safety explosives. By 1905 DuPont's experts concluded that the two most practicable and economical formulas were Carbonite and Monobel. DuPont geared up to produce Monobel and purchased Bitchel's Carbonite formula in 1906, and commenced manufacture and distribution at once[58]. DuPont continued to produce Monobel into the 1960s.

The year 1907 began the chain of events which ultimately gave rise to the wide-spread use of permissibles in coal mines throughout North America. Dr. Joseph Holmes, director of the Technologic Branch of the United States Geological Survey (USGS), became involved with studies regarding the production and availability of materials and fuels, and because coal was a major fuel, he and his colleagues decided to delve further into the causes of disasters in coal mines. Holmes had good reason to, because in 1907 the worst coal mine disaster in North America occurred in West Virginia at the Fairmont Coal Company Mines No. 6 and 8. The most likely explanation suggests that a miner did not tamp his explosive charge properly, causing a *blow-out shot* which ignited coal dust and methane in his working room[59]. Within an instant the entire mine exploded, the shock, flame, and oxygen deprivation killing an astounding 361 miners. Previously, in 1900 a blasting powder mishap ignited coal dust in a Scofield, Utah, mine, killing three-hundred twelve miners.

In an effort to address the seriousness of the issue, Holmes appointed a committee of explosives experts to study mine explosions, he obtained funding from the United States Congress, and established the Pittsburgh Testing Station under the administration of the Technologic Bureau of the USGS. The sole purpose of the Pittsburgh Testing Station was to analyze the causes and effects of blasting practices and mine explosions and fires, and the relationship between them.

After studying irrefutable links between commercial explosives and coal mine disasters, in 1909 the bureau made a general announcement to explosives manufacturers that it would begin testing formulas intended for low detonation temperature to determine if they met safety standards for use in gaseous and dusty coal mines. Those which passed certain tests were included on a list of explosives *permissible* for use. In 1909 the bureau published the first list of *Permissible Explosives*, and after that the bureau published a new list of permissibles at least once per year until 1913. Through the success of this project and others, the Technologic Branch of the USGS was reorganized as the United States Bureau of Mines in 1910, and it continued to administer the permissibles program to regulate the use of explosives in gaseous and dusty coal mines.

By 1920 every significant explosives manufacturer in the East and Midwest, and several in the West, offered a complete line of permissibles for a variety of blasting conditions. When the Bureau of Mines mandated and enforced the use of permissibles in designated gaseous and dusty mines, it created a perpetual market for them. Some mines voluntarily adopted the explosives out of safety and economic concerns. By the mid 1920s permissibles were competing heavily with blasting powder, the coal miner's favorite explosive. Until the price of permissible dynamites fell, many contract coal miners could

Table 2 Commercial Explosives Popular in the Minerals Industry: 1840 to 1960.

Explosive Type	Popular Use Date Range	Applicable Rock Density	Gases	Water Resistance	Relative Popularity	Cost	Date Invented
Ammonium Nitrate	1905-Present	Moderate-Hard	Clean	Poor	Moderate	Moderate-Low	1885
A Blasting Powder	1858-1950s	Soft	Poor	Poor	Low	Low	1858
B Blasting Powder	1858-1950s	Soft	Poor	Poor	Very High	Very Low	1858
Blasting Gelatin	1895-1930	Very Hard	Clean	Excellent	Very Low	High	1876
Extra Dynamite	1905-1960	Moderate-Hard	Clean	Moderate-Low	Very High	Moderate-Low	1879
Extra Gelatin	1905-1960	Moderate-Hard	Clean	Good	Moderate	Moderate	1879
Gelatin	1895-1960	Hard	Clean	Excellent	Very High	Moderate	1876
"Mining" Powder	1830s-1870s	Soft	Poor	Poor	Very High	Very Low	1830s
Nitrostarch	1915-Present	Soft-Hard	Clean	Good	Low	Moderate	1904
Railroad Powder	1876-1930	Soft - Moderate	Poor	Poor	High	Low	1876
Straight Dynamite	1874-1920	Moderate	Poor	Moderate-Poor	Very High	Low	1868

not, and many mining companies would not pay the $8.00 to $12.00 for a fifty pound case when a twenty-five pound keg of powder cost $1.25 to $2.00[60]. This behavior typifies the trend of saving money at the expense of safety in mines. But by the 1940s nearly all coal mining companies used permissibles in place of all other explosives.

The trend of increased use of permissibles had obvious repercussions for coal miners. They gained some peace of mind knowing their mine was less likely to explode in a chaotic hell when they shot their rounds, and there were hidden benefits too. Blasting powder and railroad powder were notorious for producing poisonous and noxious gases, while permissibles produced considerably less. Blasting powder could be found only in a few small mines by the early 1950s, when it was banished from coal fields forever, where it once was king.

CHAPTER TWO

ONE ROUND IN AND ONE ROUND OUT: THE TECHNOLOGY OF MINING AND BLASTING

"One round in and one round out" is an old expression used by American miners that succinctly summarizes the mining process. It implies that mining was basically drilling blast holes, loading them with explosives, shooting them, mucking out the shot rock, and running through the cycle again on the next shift. This chapter details the historic methods behind loading and shooting those rounds in the underground, including facets of preparing, loading, and firing explosives. It also discusses the disparity between methods recommended by mining engineers and the actual practices of miners.

To understand blasting methods and the effects of explosives in hardrock and coal mines, one must first be familiar with the basic mining process. The object of mining has always been to exploit mineral bodies containing precious metals, industrial metals, and non-metallic resources such as salt, gypsum, and coal in minimum time and at least cost.

Metallic ore bodies tended to take one of two shapes—thin and vein-like, or massive and globular. Typically gold, tungsten, and platinum tended to be deposited in *veins* while minerals such as salt, cinnabar, and some metals such as copper and iron were deposited in *massive* form. Silver manifested in both forms. Nonmetallic minerals such as colemanite, sulfur, coal, and gypsum were deposited in *beds* or *seams*. Depending on the topography over an ore body, its proximity to the ground surface, and the body's orientation, mining companies drove adits, shafts, or inclines to begin development. Once the tunnel or shaft struck the mineral body, internal workings were driven to explore the body's extent. These workings consisted of horizontal tunnels following the ore body known as *drifts*, *crosscuts* extending off the drifts,

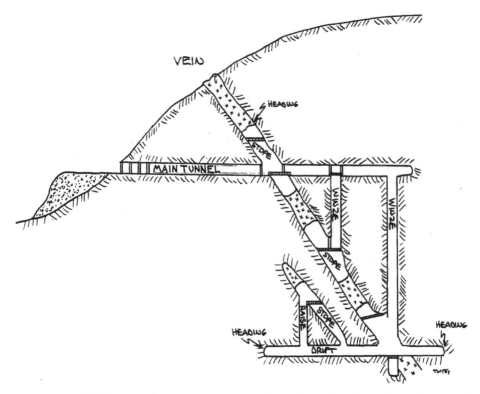

Figure 1. This illustration captures a cut-away view of a developed ore vein underground. The case presented is fairly typical in that the vein lies under a steep hillside, necessitating that miners drive a tunnel to intersect it. Once the tunnels struck the vein, they drove winzes and drifts to explore its breadth, length, and depth. After engineers and mining company directors determined the vein's extent, extraction commenced. (Source: Author)

internal shafts known as *winzes* which dropped down from the tunnel floor, and internal shafts known as *raises* which drove up.

Once engineers and geologists had determined the ore body's boundaries, miners drove drifts and raises to block out sections, and then began actual mining. The mining industry knew these caverns, chambers, and rooms created by removing the ore body as *stopes*. Those in vein-type ore bodies were often long, high, and very narrow, while those in massive ore bodies resembled vast caverns. Stopes, tunnels, and shafts were usually worked in flat faces known among hardrock miners as *headings* or *working faces*, and to coal miners as *breasts*. In coal mines the areas where coal was removed were appropriately termed *rooms*. Because of significant technological differences, we will examine drilling and blasting in coal mines separately from those practices in hardrock mines.

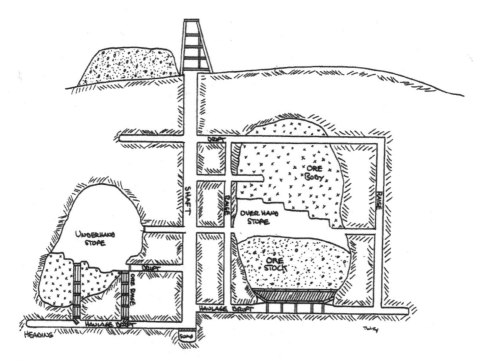

Figure 2. A cut-away view of a developed massive ore body. Because the topography overlying the ore body is flat, mining engineers sank a shaft to intersect it. By driving drifts off the shaft, company directors and engineers determined the ore body's size and extent, and blocked in for extraction. Miners worked the stope on the left with the underhand methods, from the top down, and dumping ore down into the ore raises where laborers known as chute tappers loaded it into ore cars. Miners drilled and blasted the stope on the right with overhand methods, from the bottom up, and the ore fell onto the stock pile below where laborers tapped it from underneath. (Source: Author)

Drilling a Round

Double Jacking and Widow Makers

Drilling blast-holes in the working face of a tunnel or shaft was the first step of the blasting process in any mine. Drilling methods played a key role in the speed and efficiency of mining and in the specific application of explosives. Miners did not arbitrarily drill blast-holes into the heading; rather, they universally applied special patterns designed to minimize drilling time and the quantity of explosives. As with all aspects of mining, drilling methods and hole patterns evolved and changed. Here we look at both.

Figure 3. Industrial metal ores such as silver, copper, lead, zinc, and iron occurred in massive deposits. Extraction of such ore bodies resulted in cavernous stopes which required little support in sound rock. In this scene, captured in the 1920s, several miners are drilling blast-holes in unmanageably large boulders with sinker drills, and other miners load muck into ore cars. (Courtesy E.I. DuPont de Nemours & Co.; Source: Mining & Metallurgy, Sept. 1928, p6).

Rockdrilling technology may be broken into four periods which overlap slightly in time. The first, characterized by *hand-drilling*, spanned from the time Cornish miners developed blasting in the 1700s to the 1910s. The second period ranged from when *piston-drills* began to eclipse hand drilling in the 1890s until around 1915 when *hammer-drills* became the mainstay in drilling.

The third period is defined by the use of heavy, column-mounted hammer-drills for most drilling, spanning 1915 until the 1940s. The last period, beginning in the 1940s, is defined by the replacement of the heavy *drifter* drills with hand-held *jackleg* drills.

Prior to the broad-scale introduction of mechanical rockdrills in the late 1870s, the only means of drilling blast-holes into rock was by hand with a hammer and a drill-steel. *Single, double,* and *triple jacking* all required great skill, coordination, and muscle in order to drive a hole in a reasonable amount of time.

Skilled and strong as he might be, a driller was only as effective as his tools. Drill-steels had to be made of specially tempered steel, their blades flawless, the hammerhead even, and its handle firmly attached. Before we look at hand-drilling methods, we must first become familiar with the tools of the driller's trade.

Modern historians have described the hand driller's tool assemblage as being fairly simple, consisting of a hammer, drill-steels, and a drilling-spoon. In reality, this tool assemblage was complex. Hammers used for drilling came in several sizes, each with its own application. Miners engaged in *single jacking,* the slang term for one-man drilling, typically used three-to-five pound hammers, and *double jackers* used seven-to-ten pound hammers. In some operations where money was short, miners used just about any type of sledge hammer they could obtain, but most double jackers used nine-pound hammers with elongated hexagonal heads, and a smaller, four-pound version for single jacking. As mining tools became specialized in the latter half of the nineteenth century, some tool makers began manufacturing drilling hammers featuring long, curved heads for swinging in tight quarters. But because these hammers were more expensive than conventional types, their popularity remained low.

Drill-steels, of the utmost importance to any single or double jack driller, were highly specialized tools adapted to working in the rigors of hard rock. These tools were made of hardened hexagonal or octagonal bars of high-quality steel. Their blades featured a steep angle of attack, and only experienced blacksmiths were capable of the involved forge-sharpening process. Miners always used drill-steels in graduated sets. *Starter-steels,* also known as *bull steels,* were often twelve inches long, but numerous trips to the blacksmith's forge reduced them to as short as eight inches. The rest of the steels followed in successive six-to-ten-inch increments. With each successive increase in length, a steel's blade decreased slightly in width, ensuring that it did not *fitch,* or wedge tight in the hole. Generally, drill-steels used for single jacking were no longer than three feet, because long versions had too much inertia to be overcome by the blow of a four-pound hammer. The longest steels used for double jacking were usually four to six feet in length.

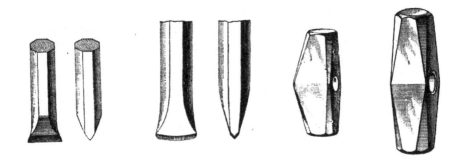

Figure 4. The line drawings illustrate typical hand drill-steels and hammer heads. Drill-steels were typically made from 1 inch hardened octagonal or hexagonal bar, and they featured blades with steep angles of attack. Most hammers used for hand-drilling ended in octagonal striking faces, and some heads were curved to facilitate swinging in tight quarters. (Source: International Correspondence Schools, 1907 A35, p3, 8, 9).

The type of rock a miner drilled determined the width of the steel's blade. If the rock was soft, a blade up to two-and-one-half inches wide could have been used, but in hard rock a narrower blade was necessary to minimize resistance. In the West where most mining was in hard rock, miners commonly used blade-widths ranging between three-fourths to one-and-one-half inches in width. Another factor affecting blade width was the type of explosive used. Dynamite packed a bigger punch per volume than blasting powder, hence holes for dynamite could afford to be three-fourths to one-and-one-quarter inches in diameter, while blasting powder required larger holes.

Drilling spoons were as important for drilling as drill-steels and hammers. Mine blacksmiths typically manufactured drilling spoons from three-eighths-inch steel rods, and they forged one end into a trough-like concavity with a scraper which miners used to scoop, drag, and ladle cuttings out of the drill-hole. The other end of the spoon was either hammered into a point, or into a flat face for tamping explosive charges. The spoon had to be slightly longer than the holes being drilled to give miners something to hold when using it.

Until the twentieth century, hardrock miners employed three methods for drilling blast-holes in rock. They knew the simplest and most popular method as *single jacking*, and it came to North America from the mines of Cornwall. Single jacking involved a miner holding a drill-steel in one hand and a three-to-five-pound hammer in the other. With each ringing strike of his hammer the miner rotated the steel an eighth to a quarter-turn until his wrist was extended, and then he slowly rotated the steel back again[1]. Hand drilling resulted in roundish but asymmetric holes, usually with a triangular shape. As a miner drilled the hole deeper, he traded the short steels for the next length in the set until all of the steels had been used. Depending on the hardness of rock and the quality of the drill's steel, a good drill-steel lasted for, perhaps, three holes before it had to be sharpened by a blacksmith[2].

Once the build-up of drill-cuttings in a drill-hole interfered with progress, the miner used a drilling spoon to remove them. For down-holes, he poured in a little water from a small bucket, made a paste out of the cuttings, and scooped it out with the drilling-spoon. Miners often scrounged vessels such as lard and peanut butter buckets, food cans, and even bottles to contain their drilling water. In horizontal holes miners spooned the cuttings out dry, while their removal in up-holes was not an issue because they simply fell out.

Double jacking, the other popular hand-drilling method, relied on a pair of miners working in close concert. The crew consisted of a *shaker* who held the drill-steel, and a *hammer swinger* who struck the steel's butt-end with a nine-pound hammer. Imagine swinging a heavy hammer for hours trying to hit the end of a one-inch diameter steel, no matter its orientation, with no more than the light of two candles, all without missing a swing and maiming your partner! When the team was drilling a horizontal hole, ideally the shaker faced the heading on bended knee with the drill steel resting on his shoulder and both hands holding its mid-section. The hammer swinger stood to one side or the other and struck the drill-steel on average every two seconds[3]. Every so often, the pair of miners would switch off on the hammer and the steel. Because hand-drilled mine workings were usually confined, the above positions could not always be practiced. In reality, the shaker and hammer swinger worked together to find the best posi-tions suited to the available space. Double jacking was best for drilling deep holes in a short time, but many confined tunnel headings, shafts, winzes, and stopes were better suited for single jacking.

Miners knew the other form of labor-intensive drilling as *triple jacking*. A shaker held the drill-steel while two hammer swingers on either side of him swung their hammers in turns with each other, striking the butt of the drill-steel. This method of drilling was rarely practiced underground as there seldom was enough space for three workers.

In relatively hard rock such as granite, quartzite, or limestone, single jacking miners were able to bore between approximately one half and one foot of hole per hour, and double jacking averaged between approximately one to one and a half feet of hole per hour[4]. With these rates, miners advanced under-ground workings slowly, and particularly hard rock only impeded progress. A moderate-size tunnel, approximately four feet wide by seven feet high

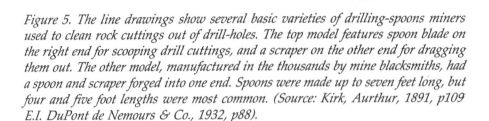

Figure 5. The line drawings show several basic varieties of drilling-spoons miners used to clean rock cuttings out of drill-holes. The top model features spoon blade on the right end for scooping drill cuttings, and a scraper on the other end for dragging them out. The other model, manufactured in the thousands by mine blacksmiths, had a spoon and scraper forged into one end. Spoons were made up to seven feet long, but four and five foot lengths were most common. (Source: Kirk, Aurthur, 1891, p109 E.I. DuPont de Nemours & Co., 1932, p88).

required nine to twenty blast holes, and a team of miners could advance it at least a foot per shift. In moderately hard rock such as rotten granite and many meta-morphic rocks, progress was at least several feet per shift.

Despite its relatively slow pace, hand-drilling had a number of advantages. Because it was labor-intensive

Figure 6. The lithograph captures two hardrock miners singlejacking blast-holes in a stope around 1900. They appear to be drilling a wedgecut hole pattern, based on fact that their drill-steels angle right. (Source: International Correspondence Schools, 1907 A35, p8).

and required little capital, anybody with physical endurance, coordination, and advanced blacksmithing skills could have practiced it, which attracted poorly-financed independent miners and prospectors. Because of these qualities, hand-drilling experienced a limited revival during the Great

Figure 7. Rarely did drilling crews have the space underground to practice triple jacking, as shown. The lithograph was probably taken from inside a massive stope, and just out of view are fresh drill-steels, drilling spoons, and a pail of water for cleaning cuttings out of the hole. (Source: International Correspondence Schools, 1907 A35 p9).

Depression, even though it was a technological throwback[5]. Another benefit of hand-drilling was its effectiveness in tight underground workings where rockdrills proved ungainly. It was also practical for blasting in mines and prospects too remote to cost-effectively haul machine-drills, drill-steels, and a compressor.

In response to the drive for ore production, mining companies began employing ever greater numbers of rockdrills in the late nineteenth century, and hand-drilling began to lose ground. This trend accelerated during the 1900s, and by 1910 hand-drilling gave way to machine-drilling, marking the beginning of a new era. The principal factors that facilitated the rise of rockdrills included an improvement in mechanical technology and a decrease in price. In the 1880s few mining companies risked employing the machines, but by the 1910s machine drills were not restricted only to the well-financed mines -many moderate-sized operations could afford to use them too. In the year 1871 Simon Ingersoll and the Rand Brothers introduced the first durable mechanical rockdrills, and within thirty years the machines changed mining methods and profoundly affected the lives of hardrock miners[6]. The first generation of rockdrills were hissing monsters powered by steam, they weighed well over 300 pounds, and they required a crew of two to operate. Subsequent generations of drills were powered by compressed air. Nearly twenty years passed before the use of rockdrills became widespread. Early models were unreliable, miners were uneducated in the ways of machine drills and operated them inefficiently, and organized labor saw them as a threat. In addition, installing an air compressor, a power source to run the compressor, and air plumbing required engineering and much capital. But by the 1890s many of these problems had been overcome; rockdrills became cheaper and reliable, and their popularity surged.

Rockdrills commonly used in the 1890s, with their steels and accessories weighed over 500 pounds. Mechanically they were simple, operating along similar principles as the slide-valve steam engine, and they had as few as five moving parts. These machines drilled holes by rapidly and repeatedly ramming a drill-steel against the rock and rotating it automatically between blows to keep the hole round.

The guts of early drills consisted of a cylinder perforated with air ducts, a reversible valve which directed the flow of air, and a piston. The chuck for the drill-steel was cast as part of the piston, and it featured a U-bolt which clamped the drill-steel's butt. When the piston was all the way back in the drill body, it tripped the valve which ducted air or steam behind it, and the expansion of either gas pushed the piston forward. When the piston returned to the rear of the drill it slid on a shaft with curved splines which rotated it slightly, in a mechanical simulation of what a shaker did during double jacking. It was this rotation that allowed drills to make neat, round holes. The drill-body moved back and forth on a sliding carriage, according to a worm-gear hand-cranked by the drill runner.

The drill runner controlled the air or steam through a throttle-valve screwed into the drill's intake manifold. A flexible iron-armored rubberized canvas hose connected to the throttle-valve via a thread-on coupling, which ultimately connected to the air lines in the mine. Occasionally these couplings managed to vibrate loose and cause mayhem. In cases where the line was charged with steam, miners were scalded, and when compressed air was used, which was the norm, if the drilling crew did not move fast enough to grab the hose, they stood to be lashed and beaten as it whipped about.

Like drill-steels used with hand-drilling, those for early mechanical drills also came in graduated sets. Starter-steels were typically twelve inches long and the longest used underground were usually eight feet. Steels used with piston-drills were made of one-inch to one-and-one-half-inch hexagonal or octagonal stock. The cutting-ends of drill-steels were known in the mining industry as *star bits* because of their star-like appearance when viewed end-on. They featured two chisel blades which crossed each other in their center, offering more cutting surface than the single-blade steels used with hand-drilling. The butt-ends of piston-drill-steels were round to fit into the chuck. Starter-steels typically had bits two-and-one-half to three inches in diameter while longer steels typically had one-and-one-quarter to one-and-one-half inch diameter bits. Diameters decreased as steel-lengths increased to prevent jamming in the drill-hole. According to numerous observations made underground, drill-holes of this era averaged four to six feet deep and between one-and-one-half to three inches in diameter.

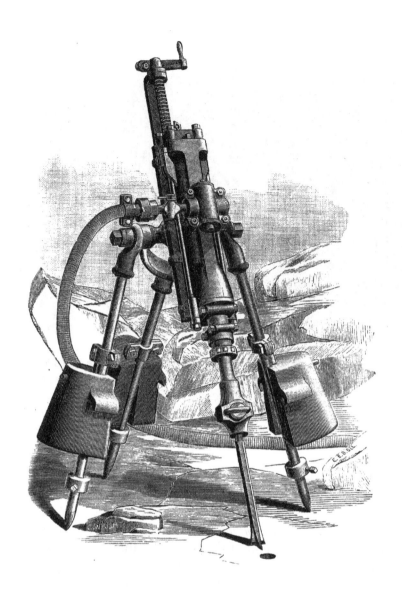

Figure 8. In 1871 Simon Ingersoll and the Rand Brothers had developed machines capable of drilling blast-holes as fast as skilled miners with hammers and hand-steels. The early rock drilling machines were unreliable, ungainly, and expensive, and as a result they remained unpopular until the 1890s when manufacturers had addressed these three issues. The end products were drills similar to the Ingersoll Rock Drill Company's Eclipse *model, shown in the lithograph. The drills' basic features consisted of a chuck with a U-bolt clamping the drill-steel, a throttle valve and air hose manifold, and a sliding carriage advanced or retracted via the worm gear. Tripod stands were rarely used in mines, where drillers favored the sturdy and stable column. (Source: Ingersoll Rock Drill Co., 1888, p2).*

Figure 9. Shaft sinking with Rand Drill Company piston drills in the mid 1880s, probably in the West. The hanging ore bucket appears to sending down sharp steels and two drilling spoons for cleaning rock cuttings out of the drill-holes. The horizontal use of columns for mounting drills during shaft sinking and for stoping in veins was standard practice. Miners usually used wood blocks as cushions between the column feet and mother rock. (Source: Rand Drill Co., 1886, p10).

In order to operate the drill, miners had to clamp it to a *saddle* which was bolted to a screw-jack stretching from floor to ceiling. The screw-jack, known by miners as a *timber jack*, a *drilling column*, or *column* was typically four to six inches in diameter with a fixed foot at the top and a threaded foot on the bottom. Some had dual adjustable feet to prevent rotating. Columns for use in tunnels were usually five to six feet tall while *stoping columns* or *stoping bars* were from two to five feet tall. On average it took between fifteen and thirty minutes to set up a column and drill, and to arrange the accessories[7].

To *drill off the column*, as miners termed operating column-mounted drills, the drill runner turned on the throttle valve, and as the machine chattered and pounded away, he slowly turned the hand-crank which screwed the drill forward and forced the steel against the rock. When the drill-steel was in the hole as far as it would go, the next length of steel was needed and the drill runner backed the drill up on its carriage and shut off the throttle valve. The chuck tender uncoupled the short steel, cleaned the drill-hole of cuttings with a drilling spoon, slid the next length of drill steel into the chuck, and tightened the shackle bolts. The team was off and running again.

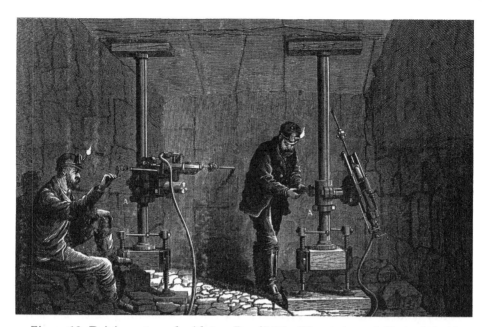

Figure 10. Driving a tunnel with two Rand Little Giant *piston drills in the mid-1880s. While most tunnels were not wide enough to accommodate two drills working together as shown by the lithograph, the scene captures the essence of what miners termed as* drifting. *Usually a crew of two operated a large piston drill mounted on a column that was tightly screwed in place between the tunnel floor and ceiling. In the scene the miner on right has shut his machine off and is preparing to reposition it to drill another hole, and the miner at left is slowly advancing his Little Giant as it bores into the heading. The oil wick lamps used by both miners suggests they are not working in the West. (Source: Rand Drill Co., 1886, p6)*

Mining companies found that the greasy machines increased ore production and sped the development of underground workings. Miners drilling off the column typically advanced tunnels and shafts three to five feet per shift, as opposed to one foot in the same rock using hand-drilling methods[8]. These improved drilling rates made roomier underground workings more affordable, allowing miners to drill and blast extra rooms, chambers, and work stations, the sizes of which were previously uncommon.

After using these machines for over a decade, people in the mining industry began to associate the drills with a very high rate of debilitating, deadly silicosis, also known as *miner's consumption* among drilling crews. The connection lay in the high quantity of dry rock flour generated during drilling, which was easily mobilized into the air by leaky air hoses, exhaust from the drills, blasting, and foot traffic. Although drill makers and mine suppliers sold their products under grand names such as *Giant, Sullivan,* and *Excelsior,* miners knew them by what they did to families—*widow makers,* a name that stuck. Yet, because these machines significantly reduced drilling time by pounding deeper, larger-diameter holes faster than hand-drilling, speeding the blasting cycle, mining companies bought them in great numbers to replace double jacking crews.

The lag time between the first employment of machine drills and the widespread acceptance of low-maintenance, cost-effective models by small and large mines alike spanned over twenty years. In 1880 only 250 machine-drills were reported at work in the entire Western States, but in the 1890s compressors and rockdrills became common components of productive mines throughout the West[9]. Widespread employment of these drills, all for expediting the blasting process, had social repercussions. The introduction of rockdrills had the unintended effect of stratifying and specializing labor in mines by creating the positions of drill runner and chuck tender, both with special skills. Mining companies also hired more blacksmiths and shop workers to keep pace with the bundles of dull steels and broken drills. Skilled drillers, shop workers, and blacksmiths became quite desirable workers among mining companies. Likewise, engineers capable of designing and installing air systems to power the drills became a prized specialty.

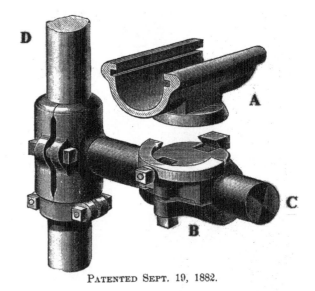

PATENTED SEPT. 19, 1882.

Figure 11. The clamp for mounting piston and large hammer drills, referred to as a saddle, *proved to be so effective that it changed little between the 1870s and 1950s. Item A is the carriage which holds the drill body, and it features a circular disk that nests into the saddle, item B. The bar labeled C is a T-bar, and it is bolted to the column, item D. (Source: Rand Drill Co., 1886, p13).*

The next significant advance in rock drilling began to take shape in 1893 when George Leyner invented the *hammer-drill* in Denver. He patented the first marketable model in 1897 and began producing an improved version in 1899[10]. Leyner's idea was a mechanical simulation of double jacking. Instead of repeatedly ramming the rock with a drill-steel as did piston-drills, Leyner's drill employed a loose piston known as a *hammer* which cycled back and forth inside the drill at tremendous speed and struck the butt-end of the drill steel, which rested loosely in the chuck. Like piston-drills, Leyner designed

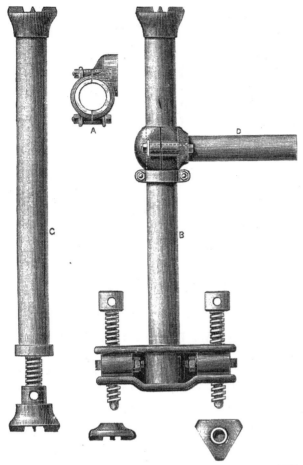

his drill for positive rotation to make round holes and to keep the drill-steel from jamming. A loose hammer moving back and forth tapping the steel was much faster and more energy efficient than moving the entire steel and chuck together, as did the piston-drill.

Leyner's drill could punch holes in rock faster than piston-drills, and crews found them easier to work with in terms of changing steels. One of the beauties of Leyner's design was the drill-steel chuck; drill-steels featured two keys on their butts which fit into receptacles in the drill's nose. All the chuck tender had to do to change steels was give the steel a twist to unlock it, replace it with a fresh steel and twist it. No longer did the chuck tender have to deal with bolts. Leyner's steels were made of one-and one-quarter-inch round stock, and they had star bits. The hammer-drill

Figure 12. The dual-foot column was the most common type of drilling column used underground prior to the 1920s. Because of a lower cost, some mining companies purchased single foot models. Up to the 1900s most columns were six inches in diameter, such as the units shown, while later types were three to four inches in diameter. Miners typically used columns from seven to five feet long for driving tunnels. They used lengths four to two feet for tight work, and dubbed the short versions stoping bars. *(Source: Ingersoll Rock Drill Co., 1888, p8).*

still required two miners to move it, it was operated in the same way as a piston-drill, it was mounted the same way, and it too had the drawback of running dry—of being a widow maker. To this regard, Leyner added a feature, which proved to be at first murder for miners, and later a true blessing. Leyner tinkered with his design and came up with a hollow drill-steel which jetted compressed air into the drill-hole while the drill was running. The air blew drill cuttings out of the hole, negating the need to scrape them out, but such action filled the miners' work area with a cloud of silica dust thicker than a London fog, which

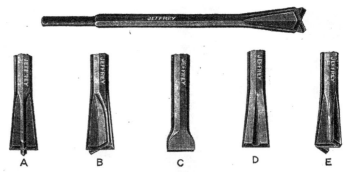

Figure 13. By the 1890s rock drill makers began offering a variety of specialized drill-steels. Steel A was used almost exclusively for drilling blast-holes underground, while the others were for specialized work. Steel B was also used underground, but less frequently. Steels C through E saw use in quarries and other types of surface work. The round end of the steel fit into the drill chuck, which clamped it with a U-bolt. (Source: Jeffrey Manufacturing Co., 1906, p12b).

accelerated silicosis. Deadly silicosis in mind, Leyner converted the air jet to water, which allayed rock dust. This technology caught on throughout the drill manufacturing industry, and by the late 1910s all new machines had the water feature and ran *wet*, curtailing the problem of silicosis. However, until the last widow makers were worn out and sent to the scrap heap, many miners continued to be sentenced to breathe rock dust.

During the fifteen years spanning 1897 to 1912, mechanical engineers instituted a number of rockdrill developments based on Leyner's machine. The prime motivation for developing new drills was to speed drilling, which ultimately made blasting more efficient. The first new drill was the *stoper*, which was a specialized light-weight hammer-drill designed to bore holes upward into ceilings. Stopers did not advance toward the rock in the same manner as hammer or piston-drills. The rear body plate featured a telescoping foot that operated under constant air pressure, pushing the drill up into the ceiling. The stoper's main significance lay in its size; it was the first self-contained drill portable and operable by one man. Because of its small size, early stopers lacked a chuck rotation mechanism. To enable the drill to bore round holes, the miner running the machine used a long handle that stuck out of its side to rock the entire machine back and forth, side to side. This odd action earned the machines the name of *wiggle tails* among miners. Because of the development of specialized drills such as the stoper, miners began differentiating the column-mounted drills as *drifters* (from using them to drive drifts), and as *Sullivans*, and *Leyners* (after two popular drill manufacturers).

Simple Water Equipment

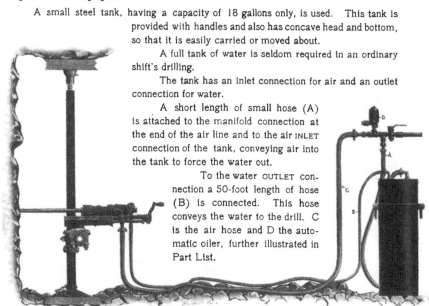

A small steel tank, having a capacity of 18 gallons only, is used. This tank is provided with handles and also has concave head and bottom, so that it is easily carried or moved about.

A full tank of water is seldom required in an ordinary shift's drilling.

The tank has an inlet connection for air and an outlet connection for water.

A short length of small hose (A) is attached to the manifold connection at the end of the air line and to the air INLET connection of the tank, conveying air into the tank to force the water out.

To the water OUTLET connection a 50-foot length of hose (B) is connected. This hose conveys the water to the drill. C is the air hose and D the automatic oiler, further illustrated in Part List.

Figure 14. In the early 1900s George Leyner began manufacturing his Rock Terrier, the first marketable hammer drill, in Denver. Leyner's drill was the first to run a jet of water through the drill-steel to flush rock cuttings from the blast-hole. The water system devised by Leyner did not become common in mines until the 1910s, and once adopted mining companies used it into the 1960s. (Source: I. George Leyner Engineering Works Co., 1906, p8).

In 1912 Ingersoll-Rand, one of North America's leading rockdrill manufacturers, developed a revolutionary hammer-drill that, decades later, would replace most types of drifter drills. Known among miners as a *plugger, shaft sinker,* or just *sinker,* the drill was basically the body of a stoper fitted with handles instead of a telescoping foot, and a mechanism for rotating the chuck. Miners used the drills to bore down-holes in the floors of shafts, stopes, and winzes. Ingersoll-Rand named their model the *Jackhammer,* which is the root of the slang name for the air tool commonly used on construction sites today. Popularity for such a rockdrill was realized almost instantly and many manufacturers muscled into the market.

By the 1910s the variety of drill types had mushroomed. So did the kinds of drill-steels that drill manufacturers offered to mines and quarries. With the take-over of Leyner's water jet technology, most steels for drifters and sinkers were hollow. Drifters continued to use one-and-one-quarter-inch keyed, round steels, and some manufacturers offered drifters that used one-and-one-quarter-inch square steels. Sinkers and stopers with rotating chucks,

Figure 15. A pair of stereotypic Western hardrock miners are driving a tunnel heading in the Colorado Consolidated Mine near Eureka, Colorado during the late 1910s. The miners are using a Leyner-Ingersoll hammer drill manufactured by Ingersoll-Rand from the late 1910s into the 1920s. In the classic pose struck by drilling crews, the miner at left clutches a wrench for loosening and tightening the heavy nuts clamping the saddle and Tbar. Note the 8-hour Wolf carbide lamp, and the two hoses, one for air and the other for water. (Source: Ingersoll-Rand Drill Co., 1921, p20).

developed in the 1920s, used square steels, one-inch hexagonal steels, or seven-eighths-inch hexagonal steels. Up until that time, however, stopers required old-fashioned cruciform steels, and as long mining companies kept the old stopers running, they continued to maintain and sharpen the old cruciform steels. In the mid-1930s drill-steels with removable, disposable bits were introduced to the market, negating having to take the whole steel to the drill-sharpener[11]. Most bits were thread-on, some were press-on, and they gained popularity through the 1940s until they replaced solid bit drill-steels in most large mines by the mid-1950s.

Hammer-drills proved to be a major improvement to miners' work environment, and to the mining industry. The fast, small drills were easier to move and operate than the old brutish piston-drills. In the eyes of the mining industry, the new drills meant greater production. With a fresh steel in the chuck, fast hammer-drills in the 1910s could drill a three-foot horizontal hole in granite in only ten minutes[12]. In only three hours of running time they were capable of punching eighteen five-foot holes in soft metamorphic rock, enough to blast a working face in a tunnel or shaft. Compare this rate with

Figure 16. Stoper drills, so named because they were well-suited to drilling up-holes in overhand stopes, surged in popularity during the 1900s. Early stopers did not automatically rotate the drill-steel like piston and hammer drills. In order to bore round holes, the operator had to slowly rotate the entire stoper back and forth 180° with the long handle on the machine's side. Note the steel spiral-wrapped air-hose and external throttle valve, typical of pre-1910 drills. The drill operator wears the classic garb of Western mines, including a candlestick hanging from his felt hat, and an extra candle in his pocket. The miner's necktie denotes him as possibly a shift boss. (Source: Cleveland Pneumatic Tool Co., ca. 1906, p3).

the seven one-foot holes drilled by a double jacking team in the same amount of time! In actuality, the total time needed to drill a face was greater than three hours. Miners had to prepare the work area and set the drill up, and when operating the machine, they had to change drill-steels, readjust the drill on its mounting for new positions, and break it down prior to loading the face. In some cases, when miners ran two drifters at once, they were able to drive a large tunnel up to six to eight feet during one shift, including setting up the drills, breaking them down, and loading the charges[13].

During the 1930s mechanical engineers introduced two more devices that changed drilling technology, heralding the beginning of what we can call the modern drilling period. The first invention was the *jackleg*, and it consisted of coupling a stoper's telescoping foot to the body of a sinker-drill via a detachable bracket. The sinker rested horizontally on the jackleg, and miners used it to perform the same function as drifters. A miner pushed the drill by its handles forward into the rock heading while opening an air-valve on the jackleg to extend its length, keeping the drill horizontal. The significance of the jackleg lay in that it formed a small drill operable by one man that drilled a wide variety of holes. Miners were able to use heavy sinker-drills on jacklegs to drill almost as fast as drifters. In addition, the machines were more versatile in their positioning at the face, and they took at most five minutes to set up[14]. Because these characteristics offered advantages over the old drifters, jacklegs had eclipsed drifter drills for all but deep drilling and work in hard rock by the 1940s.

The *auto-feed* drifter, invented in 1936, comprises the second significant advance in drilling technology[15]. The auto-feed mechanism was a small compressed air motor that advanced a drifter drill into the rock face to deepen the blast-hole while the machine operated. All a miner need do when running one of these drills was adjust the machine's mount and change steels when needed—the drill did the rest. The auto-feed proved a boon to mining companies because it reduced labor and quickened drilling, allowing miners to blast in even less time than before. Further, the auto-feed made feasible the *jumbo*, a mine car featuring up to a dozen auto-feed drifters on a boom, all operated by one or two miners. The jumbo decreased working time by simultaneously drilling many holes in a working face. The jumbo is still in use today.

Drilling technology as applied to hardrock mining follows the trend of development and refining experienced by other mining technologies. Early drilling relied on brutal hand-labor, and it spawned a breed of miners who were tough and highly skilled. In the interest of boosting production, mining companies attempted to mechanize drilling, and their efforts paid off in the long run. Early machine drills proved unreliable and expensive, but as rockdrills proved their worth and their costs fell, they became common

Figure 17. Miners operate jackleg *drills under Puget Sound in 1958. The jackleg consisted of a compressed air powered telescoping foot of a stoper bolted to the body of a handheld sinker drill. The result was a fast, inexpensive, lighweight drill manageable by a single miner. The miner at far right is operating a compressed-air dewatering pump. (Source: Hagley Museum & Library.)*

fixtures in hardrock mines. The impact that rockdrills had on labor was great, creating demand for miners with expertise in using and maintaining the machines efficiently. But the drills also accelerated the demise of many miners by creating the ideal conditions for silicosis. Through the twentieth century rockdrills became smaller, faster, safer, and more efficient. The ultimate result of the development of rockdrills between the 1870s and 1950s was an improvement in the mine as a work environment, mechanizing what was once brutal, slow hand-labor, and speeding the blasting process, which meant greater production for the minerals industry.

Driving Tunnels and Sinking Shafts: Drill-Hole Patterns

Drill-hole patterns were almost as elementary as explosives for efficient blasting in mine workings. Hole patterns, composed of several groups of holes drilled in designated locations in a working face, were a way for miners to organize drill-holes to maximize an explosion's force to break rock. All patterns required that groups of charges be fired in a sequence known among miners as a *firing order*. With almost all patterns, the first group of holes to be

Figure 18. Automatic feed mechanisms on heavy drifter drills made the jumbo *possible. Jumbos typically consisted of railcars with iron tube frames that supported multiple drifter drills. A crew of two miners were able to drill and blast a round of holes in a matter of hours with a jumbo. A high purchase price and maintenance costs limited jumbos to large mines with abundant capital. (Courtesy Hercules Inc.; Source:* Explosives Engineer, *Sept.-Oct. 1944, p237).*

fired, known as the *cut shot*, blasted out a cavity, giving succeeding groups of holes a weakness in the working face. Miners timed exactly when a particular group of charges was to detonate by cutting shorter fuses for the first shots and using progressively longer fuses for other groups. They developed several drill-hole patterns whose application was dependent on geologic conditions, drilling technology, and different sizes of headings in shafts, tunnels, and stopes. Seasoned miners and mining engineers instinctively knew which hole pattern maximized the force of a blast in a given situation, and thus minimized the backbreaking and fatiguing work they had to do in the dark underground.

One of the oldest and most commonly used drill-hole patterns that miners applied to driving tunnels and sinking shafts was the *wedge cut*. This particular pattern involved as few as two groups of holes, and as many as four groups, depending on the size of the heading. To understand this pattern, imagine looking at the blank working-face of a moderate-sized tunnel, for example four feet wide and seven feet high. The cut shot consisted of two parallel, vertical rows of holes, one to two feet apart, extending from floor to ceiling. The holes angled toward each other such that each pair was projected to meet at a distance of slightly more than three to four feet behind the rock face. If the rock was very hard, as many as five sets of opposing holes may have been drilled, while only three sets may have been used in softer rock. When this group fired, it removed a wedge of rock down the working-face's center[16].

All patterns required a set of *trimmer* holes, which defined the walls and ceiling, and they comprised the second group to be fired. Last in the firing order were the *lifter* holes. Miners drilled them at floor level and angled them down, terminating approximately twelve inches below the projected floor[17]. When the charges fired, they leveled the floor area and heaved the *muck*, as miners had termed the shot rock from the other holes, back onto a steel sheet laid by a *mucker*, a mine worker who shoveled and trammed the shot rock away in an ore car. The steel sheet was much easier to shovel rock off of than was the ragged tunnel floor.

If miners were driving a tunnel heading smaller than our example, the center cut, trimmer holes, and lifter holes were all that were needed. When they drove a moderate-sized tunnel like our example, then they would have probably drilled at least one round of *reliever holes*. If the rock was hard, they may have bored several rounds of reliever holes. Miners placed the relievers between the cut group and the trimmers, and the explosive charges blasted away the excess material[18].

Miners used the *center cut* hole pattern, also known as the *square cut*, as much as they used the wedge pattern. The square cut blasted a pyramid-shaped pocket out of the middle of the working-face[19]. To achieve this, miners drilled at least three holes that angled toward each other in the face's center area. This pattern also required reliever holes, trimmer holes, and lifter holes, and they were fired in the same order as the wedge cut. Miners drilling both by hand and by machine ubiquitously applied wedge cut and center cut patterns throughout North America. Cornish miners developed both patterns and brought them to North America by the 1850s, and they are still used in some underground operations today.

In some cases, such as when blasting out rooms, miners used the *side cut* and *bottom cut* hole patterns, which were cousins to the wedge cut. Because they removed wedges of rock along either the wall or floor, miners bored several rounds of reliever holes to blast away the bulk of the rock, and trimmers and lifters finished the remainder of the blast.

When rockdrills proliferated in hardrock mines, miners created several special hole patterns adapted to machines' limited range of mobility. The *burn cut,*

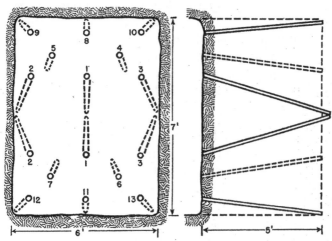

Figure 19. A horizontal wedge-cut pattern in a tunnel face. Generally miners fired holes 1, 2, and 3 in quick succession or simultaneously to remove a wedge of rock. Reliever holes 4 through 7 went off second and utilized the weakness created by the wedge cut. Trimmer holes 8, 9, and 10 fired third and they framed the ceiling, and holes 11, 12, and 13 created the new floor, and they heaved the shotrock back and onto a mucking sheet. (Source: E.I. DuPont de Nemours & Co., 1952, p264).

developed after 1920, proved to be the most popular of the special patterns. The burn cut consisted of a ring of closely spaced holes drilled in the center of a working-face, and miners loaded every other hole with explosives[20]. The empty drill-holes, between each loaded hole, served as points of weakness which facilitated breakage. The rest of the hole groups in the burn cut followed the same format as the other patterns described above. The principal advantages offered by this pattern included blasting very hard rock, and minimizing the amount of *throw*. The pattern's main disadvantage was that miners had to spend time drilling a great number of holes.

Drill-hole patterns governed the arrangement of drill-holes, but they did not determine the closeness or number of drill-holes that miners bored in a group. The rock's hardness was the primary fact that influenced miners' decisions on the numbers of holes they bored in a pattern. They understood that very hard rock such as granite, pegmatite, gneiss, diorite, quartzite, and limestone required more explosives than soft rock, and hence they spaced the drill-holes closely.

Although there were generalities for rock types, engineers and miners determined the specific number of drill-holes per hole-group and determined the overall number of holes for the entire face on a case-by-case basis. Many mining engineers stated that experienced miners familiar with a mine's geologic conditions were the best judges. Some engineers, how-ever, attempted to regulate the number of holes with formulas. One such formula suggested that two-and-one-half square-feet be allotted per drill-hole in

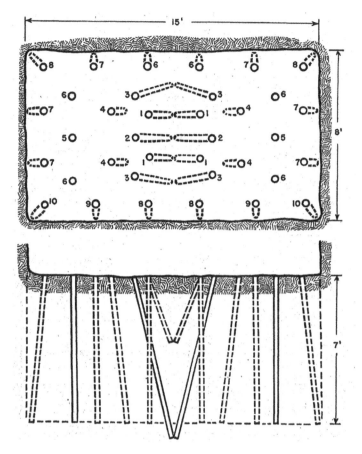

granite, four to five square-feet be allotted per hole in most meta-morphic rock, and seven to eight square-feet be allotted for sedimentary rock[21]. Real figures for square-footage allotted per hole were fairly close, suggesting that some formulas were rough but useful guides.

Drill-hole diameter also influenced the numbers of blast holes that miners bored into a working face. Hand-drilled holes tended to be shallow and small in diameter, and they contained a small volume of explosives, while machine-drilled holes

Figure 20. A vertical wedge-cut pattern for driving a haulage tunnel or for sinking a shaft. The blasting crew fired holes 1 first, which created pockets of weakness for the wedge shots. They fired holes 2, and 3 in quick succession or simultaneously to remove the wedge of rock. A first volley of reliever holes, numbered 4, went off second and utilized the weakness created by the wedge cut. A second volley of relievers went off, numbered 5 and 6, and they were followed by trimmer holes 6 then 7. Shots 8, 9, and 10 framed the corners and bottom wall. (Source: E.I. DuPont de Nemours & Co., 1952, p265).

were large in diameter and deep. This meant that a greater number of hand-drilled holes were necessary than machine-drilled holes to blast the same amount of rock. Although drill-hole length had some influence on the amount of holes packed into a working-face, it mainly determined the depth of breakage.

Blasting in Stopes

Most mining companies preferred to work ore bodies from the bottom up, because when blasted, the ore dropped into rock bins that mine workers installed under the stope, where miners easily drew it into ore cars. Conversely, working ore bodies from the top down required the installation of expensive and slow hoisting systems inside the mine to lift the ore and transfer it into ore cars. To work the ore from the bottom up, miners

Figure 21. A crew of miners is busy loading a wedge-cut hole pattern with dynamite fired by cap and fuse in the 1950s. The hanging cords are lengths of waterproof safety fuses. Note the tamping rod leaning against the timbers at right and the empty dynamite box center. (Source: E.I. DuPont de Nemours & Co., 1958, p348).

usually wedged cross-timbers, known as *stulls*, between the stope walls and they laid plank flooring across them, which served as a work platform for drilling blast-holes into the ceiling, and for readying explosive charges. Working-faces in vein stopes tended to be long, narrow, and overhead; and in the cases where a vein was quite long, it was mined from several working-faces specifically tailored to immediate geology.

Mining companies applied three popular methods for working veins from the bottom up. Miners and engineers knew the first as *overhand* stoping, they termed the second *rill* stoping, and they called the third *flatback* stoping. In overhand stoping, miners worked the ore in inverted, hanging stairsteps using mostly horizontal holes. Shooting charges in drill-holes pried the suspended blocks of ore loose, and they tumbled into the rock bins. Rill stoping was similar, except that the rectangular blocks were angled, rather than oriented horizontally[22].

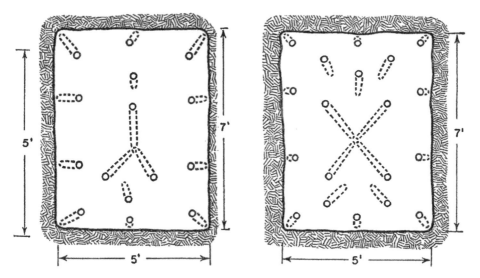

Figure 22. Diagrams for three and four-hole center-cut patterns. The four hole center-cut pattern was necessary for hard ground, which is reflected by a greater number of reliever shots adjacent to the cut holes. (Source: E.I. DuPont de Nemours & Co., 1952, p266).

Figure 23. The photo captures a drilling and blasting crew using Denver Rock Drill sinkers to bore a square-cut hole pattern in a shaft floor during the mid-1910s. Each drill operator is angling his machine toward a center point to create the initial cut round. Note how each drill has one air line each and no water lines. The man in center is either the superintendent or shift boss. (Source: Hercules Powder Co., 1918, p8).

Flatback stopes were generally confined to long and narrow ore veins, and the working face manifested as a relatively smooth, uniform surface. To shoot the face in a flatback stope, miners commonly used the side cut and the wedge cut patterns, and shooting the charges resulted in what miners termed *peeling back the face.* As implied, the side cut pattern blasted a wedge out from one of the working-face's sides, and a series of reliever shots exploding in succession across the face brought down the rest of the rock. Miners also used a *spiral pattern* to blast ore in flatback stopes, and they applied it in the Cripple Creek Mining District, among other places[23]. The firing sequence began with a cut shot in the spiral's center, and the following holes in the sequence worked their way out and around, each shot using the weakness created by its predecessor.

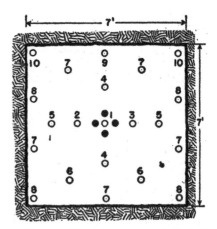

Figure 24. A burn-cut drill-hole pattern in a tunnel heading. The blank holes, represented as black dots, provided a weakness for hole number 1, which was loaded with dynamite, and it removed a pocket. The reliever holes numbered 2 through 6, and the trimmer holes numbered 7 through 10, followed in a sequence similar to the other drill-hole patterns. (Source: E.I. DuPont de Nemours & Co., 1952, p266).

Massive ore bodies presented engineers and miners with the unique problem of systematically removing immense quantities of ore while maintaining structural integrity of the cavernous stope. In sound rock which supported itself well, a mining company was free to drill, blast, and muck the maximum quantity in minimal time. In fractured, faulted, loose ground requiring support—which was every miner's and engineer's nightmare— three methods proved best in most situations.

The oldest and cheapest support system, dating back centuries, involved leaving pillars of ore in place to support the stope's ceiling. This system required no special accommodations of miners other than leaving the pillars intact. Once the bulk of ore in such a mine had been extracted, many companies, especially leasers, *robbed the pillars,* meaning the pillars were blasted out for their richness. Pillar robbing was a short-sighted attempt at obtaining ore, because it ruined the stope by removing support, which made remaining low-grade ore inaccessible. The second support system incorporated the same structural principles as leaving ore pillars, except miners erected artificial pillars consisting of timber or log cribbing filled with waste rock, and dry-laid masonry, where the ceiling was low. Building such

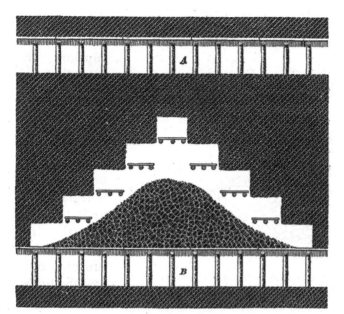

Figure 25. *The illustration shows a side-view of an overhand stope in a metalliferous vein. As miners drove the hanging rock benches backward and up, they installed timber stulls and laid plank flooring across them which served as work platforms. The blasted ore fell onto the stope floor, where laborers tapped it through chutes into ore cars in the tunnel below. Finding and working such stopes was the entire purpose of establishing and developing a mineral claim. (Source: International Textbook Company, 1899 A41, p23).*

supports permitted mining companies to remove all ore from a stope, and like the system of leaving intact ore pillars, no special accommodations had to be made for blasting, other than shoring up bad ground to preserve the drilling crews.

Square-set timbering, which was an engineering marvel, constituted the other widely used support system. Square-sets were open cells six feet long, six feet wide, and six feet high, made of heavy timbers built to fill the void created by mining out ore.

Philippe Deidesheimer, a mining engineer, developed the square set in 1860 for big mining interests working the Comstock Lode in Nevada[24]. Vernacular versions with varying cell and timber sizes were used throughout the West, several of which still stand under-ground in mines in cen-tral California, Nevada, and Utah. In many cases miners and timbermen replaced entire ore bodies with a honeycomb of square-sets. Elaborate support systems such as material-intensive square-sets could only have been constructed by well-capitalized companies possessing a talented engineer, expert timbermen, and an ample supply of timber.

Because the ground supported by square-sets was unstable, large areas of rock could not have been exposed for blasting at any one time. Instead, miners worked such ore bodies with *overhand* stoping methods from the upper stories of the timber sets. Miners laid heavy planks across the timber cells for flooring, and they drilled and blasted the portions of the ore body in front of them in giant, inverted stair steps. They drilled each face with a wedge-cut or checkerboard hole pattern, and after they shot the round timbermen quickly

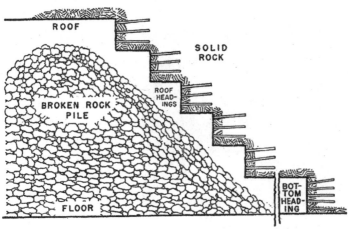

LIMESTONE

Figure 26. The illustration shows a side-view of the common horizontal drill-hole pattern used for bringing down hanging blocks of ore in overhand stopes. In some cases the mining company left the broken ore in place, like in the illustration, and it served as a work platform until the entire deposit had been drilled and blasted. This method, used in many Western gold mines, was known as shrinkage stoping. (Source: E.I. DuPont de Nemours & Co., 1952, p1317).

added square set cells to support the void. Shot rock landed either on the plank flooring where muckers shoveled it into ore cars or wheelbarrows, or it tumbled directly into underground ore bins built into the timber sets below.

One of the most popular methods of attacking massive ore bodies from the top down was underhand stoping, which tended to be a mirror-image of overhand stoping. Miners had to avoid launching valuable ore about the stope through over-zealous blasting, so they applied one of two drill-hole patterns. The first pattern consisted of several parallel rows of drill-holes bored into a bench's floor, and the other was a checkerboard pattern[25]. Provided the drilling and blasting crews did not overload the holes with explosives, when shot either in sequence or simultaneously, the blast fractured the rock while leaving it in place.

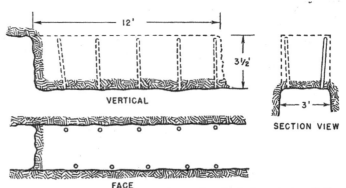

Figure 27. A drill-hole pattern for peeling back the vein in a flatback stope. The holes on the left fired first and the other holes followed in sequence. (Source: E.I. DuPont de Nemours & Co., 1952, p308).

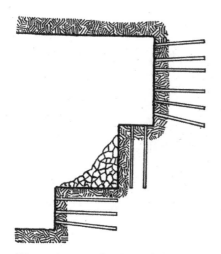

Blasting Coal Seams

Figure 28. A side view of the types of drill-hole patterns used to blast benches in underhand stopes. (Source: E.I. DuPont de Nemours & Co., 1952, p317).

Coal typically occurs in flat to gently sloping, expansive seams sandwiched between beds of base and cap rock. Mineable coal seams vary in thickness from only a few feet to as much as twenty feet, but many seemed to be just low enough to force miners to perpetually stoop. If a seam lay exposed in a hillside, a mining company could have developed it with a haulage tunnel, and if it lay under cap rock, then the company had to sink either an inclined or vertical shaft to tap it. In either case, miners drove a main drift deep into the seam for exploration and haulage, and they drilled and blasted exploratory crosscuts, often spaced 400 feet apart and up to 1,500 feet long, away from the main tunnel[26]. Secondary tunnels, parallel to the main haulage way, linked the fringes of the exploratory drifts. This network of tunnels and drifts isolated large blocks of coal, which were mined with the *room and pillar* system.

The room and pillar system, by far America's most popular method for mining coal, involved blasting the fossil fuel from a long working-face known to coal miners as the *breast*, which stood between pillars of native coal. The strength of the ceiling rock and the character of the coal determined the area of a room and the size of support pillars[27]. Strong ceiling rock coupled with dense coal pillars held up well, affording spacious rooms.

While mining coal differed technologically from mining hard rock, both types of mining shared a dependence on blasting. Like hardrock miners, coal miners had to drill blast-holes in specific patterns, load them with explosives, and fire them in sequence. To facilitate the efficient and safe blasting of coal breasts, miners had to create a structural weakness as hardrock miners did with a cut shot. Coal miners created such a weakness either with human muscle, or with explosives. The coal industry knew the labor-intensive methods as *altering the face*, and it termed the use of explosives as *shooting from the solid*. For decades miners, mining companies, and mining and explosives engineers hotly debated the efficiency and safety of the two methods. Altering the face took longer than shooting from the solid and it certainly was brutal labor, but it was much safer in gaseous and dusty coal mines because heavy concentrations of explosives in some instances ignited mine gas and coal dust. While shooting from the solid had been a culprit in touching off mine disasters, this method proved fast, translating into greater coal production. Contract miners who were paid by the ton, and some coal companies, preferred the greater tonnages produced by shooting from the solid.

We will first examine face altering practices. Coal miners used one of two means to hack cavities into coal breasts. The most popular and oldest method was *undercutting*, while *shearing* was second. With these methods miners could advance moderate-sized rooms up to five feet per day[28].

Figure 29. After a short period of mining and development, coal mines became mazes of rooms and passages navigated only by experienced miners. In the schematic diagram the white blocks are rooms that have been mined, and the black is coal left in place. The irregularities are probably areas of substandard coal. It is easy to understand that dangerous activities and practices of contract miners were nearly impossible to regulate in such mines. (Source: Peele, 1918, p718).

When undercutting, miners undermined the coal breast by hacking a horizontal channel at floor-level. Most channels were approximately twelve inches high and two to six feet deep[29]. A miner had to lie on his side in broken coal and swing a pick one-handed until his arm was tired, then switch and work with his other side. As the cut deepened, he moved closer to the breast until he was finally up to his shoulder swinging his pick. Because coal could not always support itself and threatened to crush the miner's arm, he hammered wooden wedges called *sprags* into the undercut for support, which he then had to work around. Once miners had undercut a coal breast, it had no bottom support and explosive charges easily blasted the mass loose. Coal picks tended to weigh less than nine pounds, they had sharp tines, and their handles were up to four feet long to facilitate one-handed swinging. Coal companies, large and small, well-financed or not, predominantly used the muscle of miners to gouge undercuts, but some progressive, well-financed mines switched over to machines in hopes of expediting production. The humanitarian benefits that mechanization brought to miners, who no longer had to lie on their sides and swing picks for hours, was an unintended byproduct.

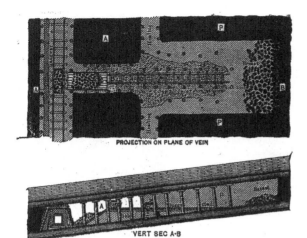

PROJECTION ON PLANE OF VEIN

VERT SEC A-B

Figure 30. Detailed plan view and elevation of a small coal room in the early stages of development. The areas labeled "P" will be mined out over time, and area labeled "A" is probably a support pillar. The short rail line in the center of the room terminates over a wood loading chute for transferring coal into train cars. It was here that the coal miner spent a good portion of his working life. (Source: Peele, 1918, p711).

Mechanical picking machines, intended for undercutting coal breasts, made their appearance in the 1870s, and only slowly became popular. One of the first types of mechanical picks, popularly known as the *Harrison Puncher*, was introduced in 1880[30]. This device was a monster which operated according to the same mechanics as the piston rockdrill used in hardrock mines, and it rested on two cast iron wheels instead of being mounted on a drilling column. To operate a puncher, two miners wrestled the machine onto a board that slanted down toward the foot of the breast, and turned the air valve on. As the bit slugged away, striking up to 200 blows per minute, the puncher slowly advanced its way into the coal breast on the wheels. Once the machine had hacked into the coal as far as the bit allowed, the team of miners pulled the puncher back, moved it over a foot or two, and started it up again. Eventually, with the mining crew's help, the puncher rammed a series of closely-spaced holes which served as an undercut.

Coal mining companies used other types of machines to create undercuts, as the technology became affordable and reliable. Coal cutters, up-scaled versions of the modern chain-saw, proved highly functional, and versions of the machines have seen use in coal mines into modern times. Horace Brown patented the first coal cutting machine in 1873, but like early rockdrills, coal cutters did not become

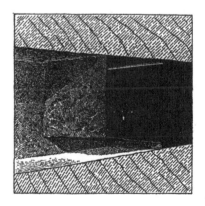

Figure 31. Block diagram of an undercut coal breast. (Source: International Correspondence Schools, 1907 A64, p61).

Figure 31A. Between the 1880s and 1910s well-financed coal companies supplied their miners with punchers *for hacking undercuts into coal seams, and for drilling low blastholes. In this scene a coal mining team undercuts a coal seam in a room with a puncher. As the machine slugged away at the coal, it slowly advanced by rolling down the inclined ramp. The undercut served to weaken the coal breast for easy blasting. Note the keg of blasting powder in the left portion of the photo. (Courtesy Hercules Inc.; Source:* Explosives Engineer *Sept. 1925, p302).*

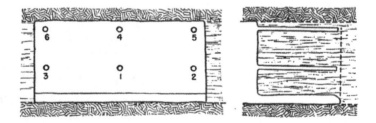

Figure 32. A common drill-hole pattern for undercut coal breasts. The miner fired charge 1, known as the buster *shot, first and it brought down most of the coal face. Charges 2 and 3 blasted down the sides, and charges 4, 5, and 6 broke the coal loose from the overlying caprock. (Source: E.I. DuPont de Nemours & Co., 1952, p300.)*

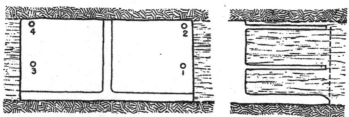

Figure 33. Diagram of a drill-hole pattern used for blasting a sheared and underct coal breast. In most cases miners did not make an undercut in addition to the shear. The miner shot the charges in the numbered sequence. Source: E.I. DuPont de Nemours & Co., 1952, p301.)

popular for some time because of high costs and mechanical unreliability. In 1877 Joseph Jeffery's company, the Jeffery Manufacturing Co. in Columbus, Ohio, introduced the first durable, mechanically reliable chain cutter[31]. The popularity of coal cutters subsequently spread throughout the coal mining industry.

Most coal cutters featured cutting bars over nine feet long, and they traveled on baby-gauge mine rail lines. To operate a coal cutter, miners laid track up to the coal breast, they moved the machine forward, and it incised an undercut. Because coal cutters were so effective, they eclipsed punchers by the 1900s, but they did not replace human muscle and sweat for undercutting until the 1920s, and manual labor was still used in some mines into the 1930s.

After undercutting, miners drilled blast-holes to bring down the breast. Because coal was relatively soft, augers with sharp cutting teeth were adequate to accomplish the task. The coal industry commonly used augers from the 1840s into the 1930s. Throughout the nineteenth century miners favored two varieties of augers: a hand-crank version much like an enlarged carpenter's drill, and a hand-cranked *breast drill*. The breast drill's auger and crank assembly fit into a concave iron strap which the miner leaned against while cranking, forcing the drill against the coal. Mechanical augers came in three basic varieties: post-mounted hand-crank types, compressed air-driven units, and electric models. All of these devices had threaded driveshafts which slowly screwed themselves into the face as they rotated the auger bit. Versions of the mechanical auger existed at least as far back as the 1860s, and why they did not replace the backbreaking breast drill is a mystery. To operate a mechanical auger, a miner had to turn the handle, while miners operating breast drills had to push against the strap with their bodies and turn the crank assembly with both arms. Both types of augers were used into the 1930s. Compressed air and electric augers made their appearance in the 1890s, but because coal mining remained labor-intensive into the 1930s, they saw only limited use.

Figure 34. Progressive mining companies began to use chain cutters for making undercuts in coal around the 1890s. While chain cutters increased production, they also cost capital which some coal companies were reluctant to spend. Early chain cutters, such as the unit illustrated, had to be transported into the mine on flat cars, taken off and placed on the room floor, and blocked in place. A compressed air motor fed the cutting bar forward on a steel frame. (Source: International Correspondence Schools, 1907 A45, p16).

Depending on the width of the breast and the density of the coal being mined, a miner could get away with drilling as few as four drill-holes. Like working-faces in hard-rock mines, blasting a breast in a coal mine required a firing order to utilize the charges effec-tively. The *buster shot* was first hole to go off in the sequence, and it was located in the center of the face. This shot blasted away the coal above the undercut, leaving a large, arched cavity. The *snubbing shot,* located above the buster shot, brought down the rest of the breast, leaving a higher arch. *Trimmers,* drilled into the upper corners of the breast, brought down the remainder of the coal. If the breast was wide, several buster shots and several snubbing shots may have been necessary, with a row of trimmers drilled along the ceiling to bring down hanging material. Some mining experts recommended using many small holes rather than a few large ones to avoid heavy concentrations of explosives[32]. Large charges caused long, hot explosions which increased the danger of igniting mine gas and coal dust. Recommended drill-hole sizes ranged from one to two inches in diameter and three to six feet deep[33].

While the coal industry greatly favored undercutting methods to create cavities in coal breasts, miners also did so with *shearing* methods. Shearing

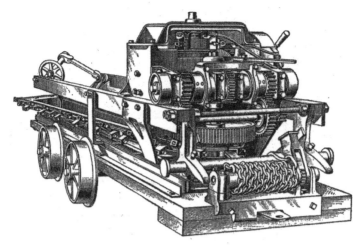

Figure 35. A chain cutter loaded onto a flat car and ready for transport into a coal room. (Source: International Correspondence Schools, 1907 A45, p17).

Figure 36. A pair of miners operate an electric chain cutter in the 1920s. Power lines have been strung across support timbers at right to the rear of the cutter, and the safety lamp served as a gas warning. (Source: Hagley Museum & Library).

involved hacking a vertical cut into the coal breast, which was mercifully easier to accomplish than undercutting, and more structurally sound. In most cases miners divided the breast into two small sub-faces by cutting a vertical trench in the middle. The hole pattern miners applied consisted of four holes, two per subface, and they angled away from the shear. The force of the exploding charges blasted the coal inward, toward the center[34].

A less popular shear method involved making a vertical cut along one of the breast's walls, rather than in the middle. Depending on the density and character of the coal, four to eight drill-holes, fired in sequence, blasted the coal into the cut.

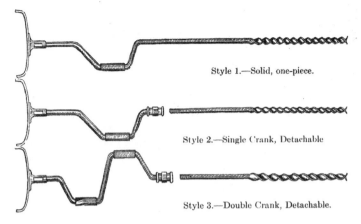

Figure 37. A sample of three basic breast augers coal miners struggled with on a daily basis to bore blast-holes. Once dull, they had to be sharpened by a blacksmith. Miners as early as the 1850s and as late as the 1930s used breast drills, until mechanized augers became universal. (Source: Dooley Brothers, 1919, p26).

Drill-holes punched near the face's center angled slightly toward the cut and the rest angled away. This arrangement blasted the coal into the cut and avoided scattering it into the room.

Shooting from the solid, also known as *shooting off the solid,* relied exclusively on explosives to bring down the breast. Contract-miners greatly favored this method because with it they produced a greater tonnage of coal in less time than with face-altering methods[35]. Because it was very explosives-intensive and required large charges, researchers had proven it to lie at the cause of a greater number of mine explo-sions and fires than face alter-ing methods. In fact, some mining companies considered shooting from the solid to be so dangerous that the practice was grounds for instant dismissal[36].

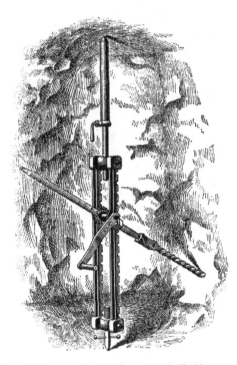

Figure 38. Handcranked post drill. (Source: International Correspondence Schools, 1907 A35, p17).

Figure 39. The grip drill, another form of coal auger, utilized an ingenious jaw that spread under force inside a short hole in the beast, gripping the hole walls. As the miner cranked the auger, the threaded shaft screwed forward into the coal. These devices were popular around the turn-of-the-century. (Source: International Correspondence Schools, 1907 A35, p20).

The drill-hole pattern used to shoot from the solid began with one hole drilled in the center of the face that was angled left or right, up or down at approximately 45°. Several mining experts attempted to govern the hole depth with formulas, one stating that a right-triangle should be formed in which the distance from the drill-hole collar to one side of the face made the longest leg, the length of the drill-hole itself formed the second-longest leg, and an imaginary line drawn from the end of the drill-hole to one side of the face made the last leg. The 90° angle of the right-triangle was at

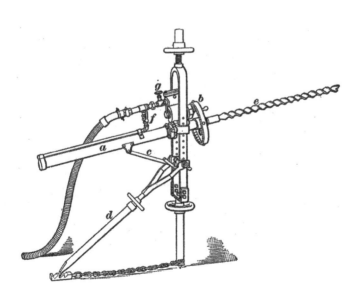

Figure 40. Compressed air powered post drill. Because these machines had a relatively high cost, they were usually used by large coal companies, and they became popular by the 1920s. (Source: International Correspondence Schools, 1907 A35, p53).

Figure 41. A miner briefly poses for the camera in a fire clay mine in the mid-1920s. Mining raw clay required methods and machinery similar to coal, and this scene could have easily been in a coal mine. Work was difficult in the low, wet seams, and it took a toll on the miners. Note the pick at bottom left and extra auger bit at lower right. (Courtesy Hercules, Inc.; Source: Explosives Engineer, *Oct. 1927, p387).*

the union of the short, imaginary line and the end of the drill-hole behind the surface of the face[37]. It is highly doubtful miners took the time to follow such a formula.

Once the first shot fired, *dependent shots* located above, below, and to both sides blasted away most of the breast. Trimmers, drilled into the upper corners and occasionally along the ceiling if the breast was wide, were the last shots fired. Drill-holes used in shooting off the solid were typically four to six feet long and two to three inches in diameter to accommodate enough explosives to break loose the coal[38].

While specific technologies for drilling and creating undercuts and shears changed, the basic blasting concepts remained the same in coal mines for nearly a century, having been transmitted through generations of coal miners. Things began to change significantly by the 1950s when giant coal min-ing machines, which greatly stepped up production, ate away at coal seams with little need for blasting.

Figure 42. In the 1890s the Ingersoll Rock Drill Company adapted its piston rock drill for use in coal mines by articulating the mount. The Radialax *featured extra chuck rotation which enabled it to cut into coal breasts quickly, and a special saddle mount allowing the operator to arc the machine and cut shear or undercut slots. However, most coal companies around the turn-of-the-century saw miners as being cheaper than drills, causing popularity for Ingersoll's labor-saving device to remain low. Note the bits and drill stems at photo bottom, and how the machine is bolted to a column. (Source: International Correspondence Schools, 1907 A45, p71).*

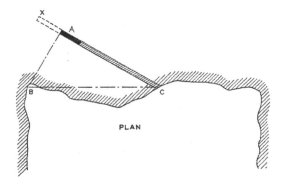

PLAN

Figure 43. Plan view of a buster shot used for shooting from the solid. Dependent shots would have been located above, below, and to both sides of the buster shot, which blew out a cavity to provide a weakness in the face. (Source: Munroe, Charles, & Hall, Clarence, 1909, p35).

Shooting a Round with Blasting Powder

By the 1840s, where we begin our look at blasting and mining in North America, explosives had become synonymous with hardrock and coal mining, and they were part of the daily existence of most of North America's miners. Prior to the introduction of dynamite in the 1870s, miners worked almost exclusively with blasting powder, and over the course of several generations they had honed the process of preparing and shooting it. Even after dynamite's popularity grew in hardrock mining, some miners continued to use blasting powder into the 1890s, and it reigned supreme in coal mines into the 1920s. By the early twentieth century many mining and explosives engineers felt that miners' application practices were so efficient and effective that they could not have been significantly improved upon[39].

Miners used a specialized set of tools for preparing blasting powder charges, for readying dynamite, and for loading both types of explosives into drillholes. Some of the tools were factory-made and could have been purchased from mine supply houses, but many tools were made by the mine blacksmith or the miner himself. Because tools cost money, many miners, especially those working under contract, improvised and used items for multiple purposes, while well-financed companies tended to supply miners with single-purpose, factory-made tools.

Because blasting caps and especially blasting powder were easily touched off by sparks and flame, mining and explosives engineers insisted that the implements that miners used to handle powder and caps, to load cartridges, and to tamp holes shut be of non-ferrous metal or wood[40]. Even though these recommendations appealed to common sense, many, probably most miners and mining companies did not follow them. Miners and mining companies often based their choice of explosives-handling tools exclusively on economics, rather than on function and safety. If the company or miner were well financed, then they were more likely to purchase the relatively costly but safer tools. Miner's needles, tamping rods, funnels for pouring powder, and powder flasks made of nonferrous metal, such as copper, brass, and bronze, cost much more than steel versions. However, in their attempts at minimizing costs, miners often chose the less expensive steel versions, even if they were more likely to create dangerous sparks[41].

The heedless practices of using potentially sparking steel implements and what was immediately at hand to handle blasting powder and dynamite can be attributed to several predominant factors[42]. Miners were often in haste to finish their shifts, or they may have been lazy or numb to danger. For example, in a coal mine in Schuylkill County, Pennsylvania, two miners were busy loading charges into a coal breast on July 3, 1923. A couple of wood tamping rods lay nearby but they were too lazy to fetch them, and instead they used a handy drill-steel to tamp the charges home, a practice forbidden by the mining company. Both miners knew they were breaking company policy, and they discussed how at some point in the future, their dangerous practices might very well cause them to "ride out in wagon." Indeed, the drill-steel threw a spark in the hole, and six cartridges of blasting powder exploded, fulfilling the prophecy of the two miners.

When working with blasting powder, miners had to carry out preparatory steps which were both fairly straightforward and very important. Failure to correctly carry out any step along the way could have resulted in a number of mishaps, some of which could have been fatal. The following includes a discussion of the general blasting processes used by hardrock and coal miners, who worked with powder underground in mines from Virginia City, Nevada, to Pennsylvania's Anthracite Fields, and from the Canadian Rockies to the Sierra Madre in Mexico.

The specific preparation and loading methods that miners followed was a function of whether they set the charges off with *safety fuse* or a *miner's squib*. Hardrock miners usually used safety fuse, while coal miners used either miners' squibs or safety fuse. A miner's squib was a gunpowder-filled paper tube one-eighth inch in diameter and four inches long with a slow-burning match end. When a miner lit the match end, it smoldered for one to two minutes until the ember reached the gunpowder and ignited it. The squib shot down the drill-hole and bumped against the blasting powder charge, setting it off. Manufacturers usually packed one hundred squibs in a paperboard box, but because the box offered little protection against being crushed and against moisture, miners used special tins to transport squibs into the mine. *Miners' squib tins* came in several varieties, but they all shared several basic physical features. First, they had external friction lids, and second, they had a capacity for at least twenty-five squibs. Models predating approximately 1910 were made with soldered construction, while some types made after 1890 had inner-rolled side seams, crimped ends, and in some cases one-piece slip-caps.

Each means of igniting powder charges had its own special preparatory requirements, which most miners seemed to master with relatively few accidents. The first step of the blasting process for both hardrock and coal miners was ascertaining which grain-size of blasting powder was best for their work.

All powder manufacturers sorted their blasting powders by grain-size and packed them into kegs. Grains considered on the small-side ranged from *FFFFFF* to *FFF*, medium-sized grains included *FF* and *F*, and larger grains ranged from *C* to *CCC* (see chart in Chapter I). This presented miners with the problem of deciding which grain-size was best for them. What constituted the best choice was relative to the type of rock they were working in. Smaller grains resulted in a quicker explosion which was best in hard rock, while larger grains caused a slower explosion, effective for blasting soft materials such as coal. Archaeological evidence in Western hardrock mining districts indicates miners used FF and FFF in hard rock while they chose F in moderately dense rock. In coal, miners typically used F, C, and occasionally CC sized grains.

Drilling blast-holes constituted the first step of an elaborate process miners referred to as "one round in, and one round out." Preparing the explosives, in this case blasting powder, was the next task they grappled with. Before miners opened their powder kegs and uncoiled their rolls of safety fuse, they had an important decision to make. How much powder did they need for an adequate blast? For some miners, gauging the most efficient quantity of powder to load into a drill-hole was a difficult task, but for experienced miners it was almost instinctual. The main goal for accurate estimation was to minimize the quantity of explosives used, which saved money and minimized gaseous byproducts, but to maximize the amount of rock blasted. In coal mines a further complication lay in the propensity of too much powder posing increased risk of igniting mine gas and coal dust, and shattering desired *lump coal* into useless, pea-sized *slack coal*.

The best means of determining how much powder to use lay within the miner—his experience, common sense, and intuition guiding his decision. The latter two qualities were at odds with mining and explosives engineers, who wanted precise quantification achieved through formulas. One formula, applied to blasting coal seams, was as follows: measure the distance in feet from the collar of the hole to its end. Multiply this distance to the 4th power. Then measure 0.4 of the diameter of the drill-hole in inches. Multiply these two quantities together, and the resulting product divided by the width of the coal seam in inches equals the most efficient amount of powder in pounds[43]. It is difficult to imagine a couple of dusty miners busily scrawling out such a formula on a piece of wrinkled fuse wrapper! Similarly, another mining engineer recommended approximately 0.06 pounds of blasting powder be used per cubic foot of coal when shooting from the solid, and that 0.01 to 0.013 pounds of blasting powder be used per cubic foot of coal when blasting an altered-face[44]. One of the more practical formulas vaguely stated "½ to 1¾ pounds of blasting powder per six foot drill-hole, depending on the density of the coal"[45]. Undoubtedly these formulas were of very limited use at best to miners who had few resources and less time to carry them out.

As is common among engineers and scientists, there were dissenting groups. Many mining engineers felt that the miner was the first, foremost, and best judge of the most efficient amount of powder. In his monograph *A Primer On Explosives for Coal Miners*, explosives engineer Charles Munroe claimed that formulas which quantify the exact amount of blasting powder were failures[46]. Mining engineer Henry Stoek expressed a similar thought in that the tonnage of coal produced per keg of powder was "purely the function of the intelligence, expertise, and judgment of the miner"[47]. In his text, *Mining Engineers' Handbook,* mining engineer Robert Peele espoused miners' ability to judge the amount of blasting powder for use in a given drill-hole. He stated:

> Black powder has so long been used in coal mining that, as regards execution only, it is considered best for this work. Most coal miners are so familiar with its use, and good miners can judge it so accurately, that excellent results are usually obtained with it[48].

If not correct, perhaps these engineers at least had a pragmatic approach.

How much powder did miners really load into drill-holes? In the coal industry, the amount depended on the method miners used to bring down the coal breast. For example, in western coal mines where miners often shot from the solid, they typically loaded their drill-holes one-half full with blasting powder[49]. In eastern mines where miners used undercutting or shearing, they used half that quantity.

Based on knowledge of how much powder miners were likely to load into a drill-hole, it is possible for the modern historian to estimate how much powder a team of miners used to shoot a working face or coal breast. Assuming fourteen ounces of blasting powder occupied one linear foot of a one-and-one-half inch drill-hole[50], a typical six foot long drill-hole held approximately two-and-two-thirds pounds. If a coal breast had two initial drill-holes and four dependent shots, miners used roughly fifteen pounds of blasting powder, which is almost a half keg. If the standard method in a mine was undercutting or shearing, as it was in most Eastern and Midwestern mines, miners often blasted with approximately half that.

Hardrock required more powder than coal did, hence hardrock miners typically loaded their drill-holes two-thirds to three-quarters full of blasting powder. In addition, drill-holes were often one-and-one-half to two inches in diameter, and four to six feet deep, which contained twenty-six ounces of blasting powder per foot[51]. A four-foot drill-hole held almost seven pounds of blasting powder. Assuming that a drill-hole pattern used to drive a hardrock mine tunnel consisted of four cut holes, seven trimmers, and four lifters, shooting one round of charges consumed approximately 100 pounds of blasting powder, equal to four kegs. These figures may be applied to the length of a tunnel, from which the total quantity of powder used to drill and blast it can be determined.

At this point in the blasting process, miners knew approximately how much powder they needed to shoot a round of drill-holes, and they had chosen an appropriate grain size. The miner faced a last preparatory step prior to loading the drill-holes. Blasting powder came in bulk form, presenting the miner with a choice of using it as it was, or wrapping it in paper cartridges. The nature of the blast-holes and the mine's environment usually, but not always, governed the miner's decision.

Putting powder in cartridges had a close relationship with miners' work environment. Of significance, cartridges kept powder dry when it was loaded into wet drill-holes and shielded it from moisture and dripping water. When powder became damp, which it did very readily, it produced an abundance of asphyxiate and toxic gases which fouled a mine's atmosphere. Regardless of these risks, some miners felt that if they poured and shot bulk powder in a damp hole fast enough, the effect of the moisture on the powder was insignificant[52]. Contract miners were notorious for taking shortcuts like this because they felt making cartridges took time away from their production. In a poorly ventilated mine, behavior such as this resulted in air quality somewhere between poor and deadly. Assembling powder cartridges also facilitated safe loading into drill-holes, because the wrappers shielded the highly flammable material from miners' flame lamps, which they used into the 1950s. Miners carelessly handling loose powder around their flame lamps put their lives in great danger. When loading powder into drill-holes that angled down, miners had a choice of bulk versus cartridges; however, drill-holes which angled upward required that miners make cartridges.

The benefits of keeping powder dry and protected in cartridges also became an issue for engineers, who of course could not agree on whether miners should do it or not. E.I. DuPont de Nemours & Company, one of North America's premier blasting powder makers, recommended always making blasting powder into cartridges, even when loading it into down-holes, because loose powder was subject to ever-present moisture and it was a safety hazard. In contrast, in the 1910s the U.S. Bureau of Mines identified a trend of accidents in which miners making cartridges by the light of their oil-wick and carbide lamps accidentally ignited the powder[53].

Miners used several methods to make cartridges. The first, advocated by explosives engineers, was somewhat materials-intensive and required special tools which miners and companies no doubt were reluctant to buy. However, with them miners produced strong, decent-quality cartridges. The other method that miners used could have been carried out very quickly, it was not materials-intensive, and it resulted in a flimsy, poor-quality cartridge. Because of its low cost and ease, it was the most popular method.

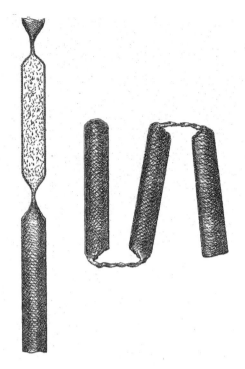

The first cartridge-making method involved several straightforward steps. First, miners coiled a sheet of paper around a pick handle, or, in some instances a wooden dowel known as a *cartridge pin*. They either glued the loose ends of paper with *blasting soap* or a like substance, or tied the open end closed with twine[54]. Blasting soap was a soft, waxy bar sold by mine-supply houses specifically for sealing cartridges, for waterproofing blasting caps, and for lubricating related equipment. Last, the miner crimped one of the tube's ends closed and sealed it. For the most part, miners used any sturdy paper to make cartridges, but some engineers espoused special water-resistant *blasting paper*, available from explosives manufacturers.

Figure 44. In late 1873 Henry M. Boies, a director for the mighty Laflin & Rand Powder Company in Pennsylvania, patented an innovative cartridge for blasting powder. Bogies invention, similar to powder-filled link sausages, would have saved miners the trouble of making their own cartridges. Why Bogies' packaging never took hold in the mining industry is a mystery. (Courtesy of Hagley Museum & Library).

Usually miners made their cartridges underground near their workings, in abandoned drifts, alcoves, and areas along side of trafficways. Apparently, miners based their choice of a work area not on safety, but instead sought a place in which they had room to work. Some historic western mines still exhibit evidence underground indicative of cartridge-making stations in such areas. Artifact assemblages, including paper used to make cartridges, unfired safety fuse cuttings, tamping rods, grease cans, and blasting powder kegs mark these sites in abandoned mines today.

After the miner completed his paper shell, he poured in blasting powder. If he planned to ignite the charge with a miner's squib then he completely filled the shell, crimped the top, and sealed it with blasting soap. If he planned on shooting the charge with safety fuse, then he inserted an end of the fuse before topping-off the shell with powder. Explosives engineers from E.I. DuPont de Nemours & Company recommended that several notches be cut into the fuse's end to allow its flame to flare, and it be tied into a knot to prevent careless blasters from accidentally yanking the fuse out. The end of the cartridge had to be tied around the protruding fuse with string or twine, and in wet environments the miners were supposed to seal the joint with blasting soap to keep water out. Although mining and explosives engineers unwaveringly insisted that blasting soap was the only acceptable sealant for waterproofing cartridges, miners also used axle grease and, in the West, candle wax dripped from their candlesticks. At the end of the process a miner held in his hand a *primed*

Figure 45. Around 1910 an anthracite miner in Pennsylvania closes a cartridge of powder he made at his work station, as had several generations of miners before him. In front of the open trunk is a powder keg on its side, to the left of the keg is a box of DuPont dynamite or railroad powder. On top of the dynamite box stands the miner's lunch pail. Note how the miner has placed his oil wick lamp several feet away for safety. (Source: Hagley Museum & Library).

cartridge, ready for loading, and made in the manner approved by engineers.

Contrary to the insistence of mining and explosives engineers, many, perhaps most hardrock and coal miners applied the quickest and cheapest method of assembling blasting powder cartridges. All a miner need do was lay out a sheet of newspaper, pour what he thought the right amount of powder was down its center, roll up the paper, and twist its ends tight[55]. Although very quick, to contract miners' delight, this method had several significant drawbacks, which had grave potential. The most significant hazard was that flimsy cartridges did not offer much moisture protection, and moist powder was one of the worst producers of poison gases. Second, there was a greater risk of loose powder leaking out of the cartridges and scattering about. Last, the twisted cartridge ends were not conducive for tight tamping, which softened the explosion and broke less hard rock. Several blasters' supply houses and E.I. DuPont de Nemours & Company began to sell *tamping bags* as an alternative to the flimsy, twisted paper cartridges, in the 1910s. These were ready-made paper cartridge shells eight inches long. Tamping bags sold successfully through the 1950s. However, as long as independent miners blasted with powder, they continued to roll their charges in paper.

Loading the Round: The Day is Almost Done

Miners ignited blasting powder charges with either miners' squibs or safety fuse, and each device required specific loading practices. Generally, hardrock and many coal miners used safety fuse, while the use of squibs was restricted to coal mines. A miner's squib consisted of a short slow-burning fuse made of paper impregnated with gunpowder, and a long, thin paper tube filled with high-grade gunpowder. A miner lit the match-end which burned down to the powder tube between approximately one and two minutes. The powder in the tube flared and the squib shot down the drill-hole like a bottle rocket where it bumped up against the blasting powder charge and ignited it[56].

S.H. Daddow filed the first patent for a squib, which the U.S. Patent Office granted him on March 17, 1874[57], however, squibs already existed by this date. Beginning in the 1880s, coal miners used them in increasing numbers, until they reached a zenith between approximately 1890 and 1910. Out of economy and familiarity some coal miners continued to use squibs into the 1930s, albeit in dwindling numbers.

Squibs possessed numerous disadvantages, the most notable of which was the fixed length of the match-end, and it burned in only one to two minutes. Fuses of such short lengths made it impossible to fire the number of holes typically used to shoot the working face in a hardrock mine, because they offered little escape time. Another of the squib's drawbacks was that it had to rest at the mouth of a guide-hole leading to the powder charge, hence it could not be used in steep up- or down-holes without falling out of place. Another problem lay in the exposed match-end, which was subject to the degradational effects of humidity, and because it was completely exposed when placed in a drill-hole, a drop of water could easily extinguish the flame. Imagine a poor miner who has lit three squibs with two others to go. Just then, a drop of water puts the first out. Does he finish lighting the round, which he knows will not fire in the correct order, or does he try to substitute another, fresh squib while the others smoldered away? Last, a squib's flames had been proven to ignite mine gas, which was the foundation for at least one mine disaster[58]. However, because squibs were cheaper per round than safety fuse, they were popular among coal miners, some of the lowest paid in the minerals

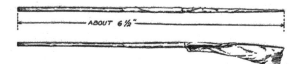

Figure 46. Side and plan views of a miner's squib. Used primarily for coal mining, a miner lit the match-end, right, which burned to and ignited a paper tube of high-grade gunpowder. The squib shot like a bottle rocket into the blasting powder charge, igniting it. (Source: E.I. DuPont de Nemours & Co., 1920, p16).

COPPER NEEDLES.

HARDSOCG   DRAWN FROM THE EYE, SPECIAL LONG SLIM POINT.

Figure 47. Miner's needles consisted of pointed rods usually made of brass, bronze, or copper, up to seven feet long. Needles were necessary to provide a passage through the stemming material in drill-holes for a miner's squib. (The Martin Hardsocg Company, 1912, p43).

industry, for approximately forty years.

Loading the charges of blasting powder was the last step taken by miners prior to shooting the working face. Miners loaded bulk powder differently from cartridges, and they loaded charges to be fired with safety fuse

Figure 48. Cut-away view of a loaded drill-hole with a miner's needle in place. After packing the tamping material into the hole, the miner withdrew the needle and set a squib at the mouth of the passage. Source: E.I. DuPont de Nemours & Co., 1920, p16).

differently from those to be fired with miners' squibs. The methods that miners used were universal in both in hardrock and coal mining industries. But before miners did any loading, they first checked to ensure the drill-holes had relatively smooth walls so that the paper cartridges would not be torn open, causing the loose powder to spill out. In most cases the rock was solid enough, but if the drill-holes were in fact *ragged*, their walls broken and sharp, ideally miners smeared them with a fine-grained material such as clay to smooth them out in what they termed *bulling the hole*. In actuality many miners did not make the effort, considering bulling an unnecessary, time-consuming hindrance.

After the drilling and blasting crew had pushed all of the cartridges into the drill-holes and loaded the primers, one of the miners sealed the remainder of the hole by packing in *stemming* material with a *tamping rod*. The miner fed bits of stemming into the hole and rammed it against the explosive charge as firmly as his muscles would allow, ensuring that the force of the explosion would not escape out the mouth of the hole. In most cases coal miners gave careful consideration not to use coal dust because the blast might have ignited it, possibly resulting in a mine disaster. Miners used any dampened, fine-grained material for stemming, such as drill-cuttings and clay. Miners used tamping rods to pack, or *tamp*, explosive charges and capping material into drill-holes. Tamping rods were often made of non-sparking material, and

their sizes were usually less than one inch in diameter and approximately six feet long. Most miners working with powder used broom handles or steel drilling spoons with wood blocks on their ends, while some miners working with dynamite also used drill-steels. Miners with disposable income could have purchased factory-made bronze and copper-tipped steel models. In addition to the above types of tamping rods, miners or companies with limited incomes used alternative rods fashioned from salvaged materials including trimmed pine branches, iron pipes with wood dowels jammed into their ends, and iron bars.

Figure 49. Lithograph, published 1869, captures typical activity at a miner's priming station. One miner prepares to fire his shots via squibs by lubricating a blasting barrel with a bar of cartridge soap. Miners used blasting barrels to provide a dry unobstructed pathway for squibs through stemming material. Note the roll of blasting paper for making cartridges located on chest's edge. (Courtesy of Hagley Museum & Library; Source: Harper's Weekly *Sept. 11, 1869).*

Once all of the powder charges were packed down and the drill-holes sealed with stemming, the task of loading the round was complete.

Thomas Wheatley in Wilkesbarre, Pennsylvania, patented a unique tamping method for better blasts in some coal seams in 1878, which

Figure 50. While most miners used dowels and broom handles for tamping rods, mine supply houses offered well-made copper or brass-tipped steel models. (The Martin Hardsocg Company, 1912, p413).

came to be known as the *air shot*. For the air shot, miners pushed cartridges into the drill-holes as described, but before tamping in the stemming material, they pushed in wads of paper or straw to create an air pocket[59]. The air pocket damped the initial shock of the explosion and spread the force out laterally, maximizing production of lump coal and minimizing slack-coal.

Despite a unanimous use of stemming material in the mining industry, as late as the early twentieth century a few miners of the "old school," often in coal fields worked from an early time, thought stemming was unnecessary. They blasted drill-holes wide-open or at best left their tamping rods in the holes as plugs[60]. This resulted in a *blow out* shot where the explosion's force, heat, and flame spewed out of the hole and little material was blasted. Because combustion was not properly contained, the powder did not burn at its optimal temperature, which was approximately 2,000° F, producing poisonous gases. Another problem associated with blow-out shots was that they easily ignited mine gas and coal dust, when present. As an example, coal mining engineers attributed a blow-out shot to igniting a hell in West Virginia's Monongah Mines No. 6 and 8, where 361 miners were consumed by fire, smoke, and lack of oxygen in 1907, many leaving families without husbands and incomes[61].

Figure 51. Front view and top view of the tip of a special copper-tipped tamping rod featuring a special cleft to accommodate a miner's needle, as it lay in a drill-hole. Such tamping rods were rarely used, and they were manufactured between the 1890s and 1920s. (Source: International Corre-spondence Schools, 1907 A36, p139).

When miners shot their rounds with safety fuse, they tamped stemming material firmly around the line of fuse trailing out of the drill-hole. But preparing a round for shooting with a squib required several additional steps. A miner stuck the last cartridge, which was sealed on both ends, onto the end of a *miner's needle*, and slid them together into the hole as a unit. A miner's needle was a rod of non-sparking metal, such as brass, bronze, or copper, one-quarter to three-eighths inches in diameter, four to seven feet long, with a sharp point on one end and a handle on the other. Some needles were made at mine

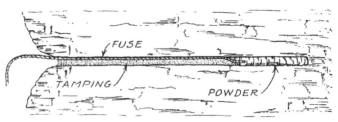

Figure 52. Properly loaded charges primed with safety fuse consisted of a powder cartridge and stemming material well-tamped in the drill-hole, and a straight fuse of safe length unmarred by tears or kinks. (Source: E.I. DuPont de Nemours & Co., 1 917a, p22).

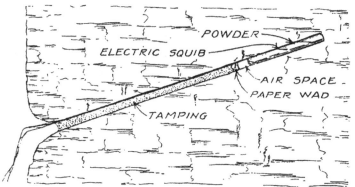

BORE HOLE IN COAL SHOWING AIR SPACE BETWEEN CHARGE AND TAMPING

Figure 53. Cut-away view of a drill-hole in coal loaded for an air-shot. (Source: E.I. DuPont de Nemours & Co., 1917a, p22).

workshops out of rod stock while others were factory-made with iron handles riveted on. The miner left the impaled cartridge and needle lying together in the drill-hole and tamped stemming

Figure 54. In this classic coal mining scene, a miner is engaged in loading permissible dynamite cartridges in a neatly square room. The miner has primed the cartridge he is loading with a blasting cap crimped to five feet of safety fuse. Judging by length of the copper-tipped iron tamping rod and drill-hole scraper at right, the miner has bored fairly deep holes. (Courtesy Hercules Inc.; Source: Explosives Engineer Oct. 1923, p220).

material around the needle, as he would if it were safety fuse.

After firmly tamping stemming around the needle, the miner withdrew the needle from the hole in a twisting action, leaving a smooth, small-diameter passage for the squib to travel down. In wet or ragged holes miners substituted a *blasting barrel*, which was merely brass tubing, to provide a dry, unobstructed pathway[62].

When possible, miners, especially those on a contract basis, favored using bulk powder because it expedited loading. A miner simply poured powder from a flask or from the keg through a funnel, filling a portion of the drill-hole. He inserted a length of safety fuse into the hole and poured the rest of the powder around it. Afterward, he tamped stemming over the powder, sealing it. If a miner chose to shoot the round with a squib, he poured powder into the drill-hole, set a miner's needle in it, tamped stemming around the needle, extracted it, and placed a squib in the needle-hole's mouth.

Electric Squibs

In the 1910s blasting with electricity gained considerable momentum in underground and open pit mining, and in quarrying. The hingepin permitting this practice was the electric blasting cap which detonated when a current passed through a filament, which was simply a resistor that heated up and set off the cap's filling of mercury fulminate. In the late 1910s E.I DuPont de Nemours & Company began selling a spin-off, the electric squib, which flared up to ignite powder, instead of exploding. By connecting many electric squibs to one blasting machine, it was possible to shoot a number of holes with near-perfect simultaneity, or by wiring in a delay circuit, a well controlled firing order. The new possibilities that electric squibs brought were perfect for some coal mines.

Figure 55. Electric squib, actual size. (Source: E.I. DuPont de Nemours & Co., 1932, p36).

Miners primed cartridges with electric squibs as they did with safety fuse, and they loaded them into drill-holes in the same way. When assembling cartridges, miners inserted a squib, poured in powder, tied the end of the paper tube closed, and secured the wire around the cartridge. They gently pushed the primed cartridge into the drill-hole, and tamped stemming material firmly around the wires[63]. Once miners finished sealing the holes, they connected the squib's lead wires to spools of blasting wire which they trailed out to the nearest place of safety. There the wires were hooked up to a blasting machine or a battery box, completing the circuit.

Figure 56. Often explosives makers demonstrated proper handling and preparation of their products in their advertisements. In the 1930s fuse maker Ensign Bickford ran an ad showing a cut-away view of a pellet powder cartridge primed with slit safety fuse. (Courtesy Hercules Inc.; Source: Explosives Engineer *May 1934, p151).*

Any operation which used electric squibs, and there were but few, was most likely a large, well financed mining company. Capital was necessary to purchase the equipment and electric squibs, which cost more than match-lit versions.

Pellet Powder

Pellet Powder, a compressed form of blasting powder, came late to the explosives market. Although known among English miners earlier, North American explosives makers, probably DuPont, introduced it to the United States in the mid-1920s, possibly as an attempt to bolster sagging blasting powder sales. Powder in this form was incredibly easy and time-saving to use, hence its popularity soared. By the 1930s pellet powder could be found in mines, quarries, civil engineering projects, on farms, and any other place where blasting powder had been used up to that time. However, in hardrock mines dynamite continued to reign, excluding pellet powder from the industry. Still, pellet powder saw popularity into the 1950s.

Explosives manufacturers produced pellet powder by packing four two-inch pellets in waxed paper cartridges eight inches long. Each cartridge had a three-eighths inch hole in its center for inserting safety fuse or an electric squib[64]. To prime a cartridge, a miner merely pierced its end and inserted either a fuse or an electric squib, and he lashed the whole thing together. Pellet powder came in several different densities which simulated the explosive behavior of blasting powder's different granulations. The main advantages of pellet powder were that it was very convenient to use, and it was safer than conventional powder because there were no loose grains.

Spitting the Fuse

Spitting the fuse is an old miners' expression which described the act of lighting the fuse of an explosive charge. Miners derived the term from the behavior of safety fuse upon being lit—a startling gust of flame and gas shot out its open end in a hiss. Spitting the fuses for a round of powder charges was more complex than many people realize. The miner responsible for blasting had to ensure that all of the fuses were in good shape, he confirmed the firing order, and he kept a cool head and a slow hand while lighting each and every fuse. As a hardrock miner prepared to light the round, what he saw was a working face with over a dozen fuses hanging out of the muddy blotches marking the sealed drill-holes. Early in the twentieth century a representative for E.I. DuPont de Nemours & Company reported that some contract miners in eastern coal mines who supplied their own fuse intentionally cut dangerously short lengths in hopes of saving money[65].

Because contract miners had to pay for their blasting supplies out of pocket, they had a propensity to minimize costs and short themselves in the process, in this case by using short fuses to save money.

Safety fuse came in either individually-wrapped fifty foot rolls or on 3,000-foot capacity steel spools—hence a miner could have cut any practical length. By the 1880s leading powder manufacturers offered fuse sheathed with resin, twine, jute, gutta percha, or flexible waterproof tape. Each skin had different water-resistance, kinking resistance, and durability against abrasion and laceration. Most safety fuse manufactured before the 1920s burned at a rate of two feet per minute while fuse used later burned at a rate of one foot per forty seconds[66]. This meant that six feet of fuse gave miners only three minutes to light the entire round and retire to a sheltered area. Anything shorter than six feet, as some contract miners used, was downright life-threatening, yet stupidity persisted. To prepare a fuse for easy and sure ignition, a miner cut a very short bit off its end and sliced it horizontally to fully expose the powder train.

As an example of what can happen with using short fuses, a miner who was loading drill-holes found he had only twelve feet of fuse left with which to set off six charges of dynamite. He boasted to the other miners that he could detonate all of the charges using only two feet of fuse each, despite the miners' protests. An experienced miner would have flatly refused to proceed; however, the reckless miner proceeded to light the fuses in order with the flame of his carbide lamp (for he did not leave enough fuse to make a spitter). Unfortunately for him, when he had lit half of the fuses, a drop of water put one of them out. Quickly he began cutting off the water-spoiled end and as he relit it the other charges detonated. The miner was hurled down the drift

Figure 57. By the 1880s explosives makers offered at least five varieties of safety fuse, each for use in different conditions. Hemp, at bottom, cost least but offered the powder train the least protection against kinking, abrasion, and most importantly, against moisture. Gutta percha, at top, cost the most but performed best. The illustration accurately depicts the appearance of the sheathing associated with each fuse type. (Source: Kirk, Aurthur, 1891, p25).

twenty feet and lay in a crumpled mass when his partners, who heard the explosion but no footsteps preceding it, found him. They were astonished almost as much as the miner himself to find a good number of cuts and scrapes, but no broken bones or lost ears and eyes[67]. By using the short fuses, the lucky miner left himself between only one minute and eighty seconds to light all six and flee to safety, with no margin for error.

Figure 58. Explosives makers offered safety fuse in quantities of 3000 feet wound on iron spools for large operations by the 1910s. Until the 1930s the spools were shipped in wood boxes, and afterward in fiberboard boxes. (Source: E.I. DuPont de Nemours & Co., 1932, p46).

In western hardrock and coal mines, miners typically used approx-imately six to eight feet of fuse per charge[68]. The exact lengths of the fuses were cropped so that those leading to the cut-holes were shortest, the reliever shots were the second shortest, the trimmers third, and the lifters longest. This assured the firing order, because the short fuses burned in less time. A firing order of sorts was achieved with squibs when a miner lit their match-ends in sequence, although it was not as controlled as with safety fuse.

The most favored method of spitting the fuse was with a rat tail or spitter, examples of

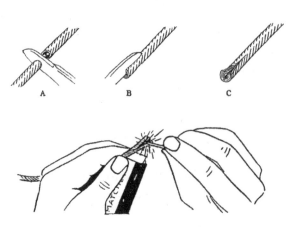

Figure 59. Lighting safety fuse required special preparation and aggressive use of flame. Simply holding a burning match or other flame to the fuse end was not always sufficient for sure, smoke-free ignition. First the miner had to cut off the old tip and slit the fresh fuse to expose the powder train. Second, he had to embed a flaring match-head into the powder train, when not using a spitter. (Source: E.I. DuPont de Nemours & Co., 1932, p46).

which may be encountered in historic hardrock mines today. The gust produced by a fuse when lit was strong enough to extinguish most small flames, thus miners never used their candlesticks or oil wick lamps to light a fuse, lest they be left in darkness immediately prior to a blast. A spitter was a short section of safety fuse between six and ten inches long, with a number of notches cut to the powder core equal to the number of fuses to be lit. Each time flame spurted from a notch in the spitter, the miner would use it to light a particular fuse. If there was one too many or one too few notches in the spitter after all of the fuses had been lit, then the miner knew that something was dreadfully wrong, and that somewhere he miscounted. Although time was of the essence in lighting a round, cool accuracy was everything. If there were many fuses to light, two miners worked together and lit opposing fuses, one of them calling out which hole the pair was lighting. It is important to note that when a miner shot a round he lit the fuses in the firing order. After the fuses were spitting and smoking away, the miner yelled the traditional "fire in the hole!" and calmly retreated to the safety of a crossdrift where his partner waited.

After the charges were lit, all was quiet while the miners listened for the report of each hole. A good mining crew was able to tell by the muffled thumps of each charge if all went off as planned. Blasting was the last task a crew did at the end of their shift because poisonous and inert gases choked the heading for several hours or more, if there was no forced-air ventilation. Once enough oxygen returned to the workings to sustain life and candle flames, the next shift arrived and began another cycle of one round in and one round out.

Blasting with Dynamite

The year 1868 held one of the most momentous events in American history, one which forever changed the mining and construction industries. In that year the Giant Powder Company, organized in San Francisco in 1867, introduced dynamite to North America. Although hardrock miners did not know it, this was a fateful year for them too, because dynamite dramatically impacted their work environment. During the 1870s and 1880s, miners had to learn to handle dynamite in harsh mountain and desert climates without ruining the product or killing themselves, and they were forced to develop efficient and effective blasting processes in the dark underground. For the most part, miners took the processes they traditionally employed for blasting powder and adapted them to the requirements of dynamite. By the 1890s miners had developed consistent preparing and loading practices, but it took miners several decades to refine the process into something efficient and safe.

As was the case with blasting powder, using dynamite involved fairly straight-forward but very important steps, which when improperly done resulted in miners underground loosing hands, eyes, and occasionally their lives. Miners did not indiscriminately shove dynamite cartridges into drill-holes, light the fuses and run. Before miners touched any explosives, the miner or mining company had to ascertain the best high explosive product for their work, and they also had to select the best strength of the explosive and the most efficient size of cartridge. Once this was accomplished, the dynamite cartridges had to be primed and readied, after which they could be loaded. In this section we look at the processes miners typically went through when blasting with dynamite, and the tools they used in the process.

Even before miners or prospectors broke ground, they had to address issues regarding the use of dynamite for blasting. First, miners and mining companies faced the problem of choosing one of several types of explosive products for their work. In the first decade and a half of dynamite's existence the choice was simple; blasting powder, nitroglycerine, and straight dynamite were the only three explosives widely available to the North American mining industry. By the early 1880s many miners found dynamite to be best in hard rock while blasting powder was most effective in soft rock. With the popularization of dynamite, nitroglycerine quickly dropped from the scene because of its reputation for being extremely temperamental. By the mid-1880s the diversity of dynamite products began expanding, making the decision of which explosive to use more difficult. Between the mid-1880s and 1900 mine supply agencies in hardrock mining districts typically carried or could order *straight dynamite, dynamite extra, gelatin, gelatin extra, ammonium nitrate, railroad powder, blasting powder,* and in rare cases *chlorate/nitroglycerine-*based dynamite.

The variety of explosives continued to expand between the 1900s and 1910s, and in coal fields, add to this roster a variety of permissible dynamites. How did a miner or mining company decide what to go with? Part of the answer depended on the performance of the explosive and how it worked regarding blasting conditions, the type of rock, and air circulation underground. But the greatest criterion miners and mining companies based their choice on was how much money they were willing to spend.

In fact, even after dynamite had superseded blasting powder in most hardrock mining districts, the poorest mining operations did not choose dynamite at all, rather they continued to buy blasting powder or railroad powder, which per pound cost less than half that of dynamite. The price of blasting powder in the inland West between around 1905 and 1930 ran as low as $1.25 and it averaged $1.75 per twenty-five pound keg, while a twenty-five pound box of dynamite cost on average $3.50[69]. During the 1890s the cost of dynamite had fallen to as low as $2.50 per twenty-five pound box as a result of a trade war among San Francisco Bay Area-based dynamite makers[70]. But by far most mining companies, well-capitalized or low-budget, chose tried-and-true straight dynamite. Although it was not well suited for work under water, and although it produced relatively foul byproduct gases, it became the industry standard because it worked well-enough in most conditions, and above all it was the cheapest type of dynamite.

However, some well-financed mining companies could afford to select the most effective explosive for their work, even if it cost more than straight dynamite. Generally gelatin cost up to a dollar more per fifty-pound box, the standard shipping weight, than did straight dynamite[71]. Companies of this caliber supplied gelatin or gelatin extra to miners working in wet conditions, dynamite extra or ammonium nitrate if the ventilation poor, and ammonium nitrate or nitrostarch if storage conditions were below freezing. Gelatin and gelatin extra had the economic advantage of blasting more hard rock than straight dynamite, hence in some cases blasting with gelatin was cheaper in the long run, with the bonus of cleaner byproduct gases. Between approximately 1910 and 1915 companies began to favor gelatin and extra formulas over straight dynamite.

In the mid-1920s the number of specialty blasting products mushroomed, complicating miners' decisions regarding which dynamite they chose. In addition to general purpose dynamites for hardrock mining, explosives manufacturers introduced variants formulated for gypsum mining, salt mining, quarrying, blasting friable material, blasting coal, and subfreezing temperatures. Like mining companies operating prior to the 1920s, many later companies based their choice of dynamite on price, while a few outfits chose specialty explosives for their performances.

By the 1930s when economic conditions became grim for the minerals industry, the cost of running a mine was of prime concern, and the price of dynamite was a focal point of cutting expenses. In response to the dismal economic climate, explosives manufacturers introduced lines of dynamite composed mostly of ammonium nitrate mixed with a little nitroglycerine. The new formulas were very low in price and excellent underground. DuPont's versions were *Special Gelatin* and *Gelex*, the Hercules Powder Company's were *Herculite*, *Hercomite*, and *Gelamite*, the Atlas Powder Company named its dynamites *RXL*, *Apcodyn*, *Amodyn*, and *Gelodyn*, and the Illinois Powder Manufacturing Company's included *Powdertol* and *Ajax*. Naturally, mining companies gravitated toward the cheaper dynamites, but still some hardrock miners continued to use explosives appropriate for specific blasting conditions, despite their slightly higher cost.

The 1910s saw the rise of permissible dynamites for coal mining, presenting coal miners with the same issues of product choice. The coal miner, like the hardrock miner, based much of his decision on cost, but he considered performance with almost equal weight. The coal he mined was of a particular character, including qualities such as its type, density, friability, and water saturation. These factors required permissible dynamites of specific strengths, quickness, and water-resistance. For example, a fast permissible dynamite appropriate for hard anthracite shattered soft bituminous coal into worthless fragments, and slow permissibles could not have adequately blasted hard coal. A few old-time coal miners still preferred railroad powder because it was very cheap and performed well, but after the 1910s many miners, mine officials, and engineers frowned on it because it was one of the worst explosives for producing foul gases[72]. The variety of permissibles for blasting coal expanded during the 1920s, giving coal companies a wide selection to choose from. Coal companies, like hardrock mining outfits, continued to buy the cheapest dynamite, but of a type suited for the work.

In addition to selecting an appropriate type of dynamite, mining companies had to supply their miners with the most efficient *percentage strength*. During the late 1860s and early 1870s, dynamite manufacturers found that varying the amounts of the explosive ingredients in their dynamites affected the explosion's strength, duration, and quickness. Straight dynamite with a large percentage of nitroglycerine exploded much faster and with more of a shattering effect than dynamite with only a little nitroglycerine. The same held true for dynamites made with explosive ingredients other than nitroglycerine. The net effect was similar to the control afforded by blasting powder's different grain sizes: dynamite with a high percentage of explosive ingredient was better in hard rock, while dynamite with less explosive ingredient produced a slower, heaving explosion, which was best in soft rock.

As with blasting powder, the performance of dynamite had to be adapted to the variety of rock conditions encountered in mining. Explosives manufacturers found that by adjusting the amount of nitroglycerine in each cartridge, dynamite's shattering effect could be dampened or amplified. Further, they determined that if they absorbed the nitroglycerine in a highly combustible base, the base's compounds transformed into gas, which added an extra push to the explosion. Dynamite with less nitroglycerine and a combustible base was best in softer rock because its explosion was more heaving, while hard, dense rock required the powerful jolt of energy provided by a high-percentage of nitroglycerine, followed by a moderate amount of heaving given by high-nitroglycerine dynamite.

To meet the varying rock types that miners encountered underground, before 1910 explosives manufacturers offered three principle grades of dynamite consisting of 60%, 40%, and 30% nitroglycerine, which they labeled as *No. 1*, *No. 2*, and *No. 3*, respectively. In addition, manufacturers offered *specials* consisting of intermediate percentages of nitroglycerine. After the year 1900 explosives makers specified the actual percentage of nitroglycerine in dynamite, which by that time could have been ordered in 5% increments between 10% and 70% of total cartridge volume.

The percentage strength of dynamite chosen by miners and mining companies was based more on performance and less on economics. Most mining companies supplied their miners with 60% strength for blasting very hard rock because it had a quick, shattering explosion. But it was used only when necessary, because it cost more than a low-percentage dynamite. For soft, friable rock miners favored 20% to 30% dynamite, because the explosion was long and heaving, and they used 40% strength dynamite for nearly everything else[73].

Dynamite made in the factory was a light-brown to gray, oily substance which manufacturers packed into waxed paper cartridges. Explosives manufacturers offered cartridges in a variety of diameters and in several lengths, which presented miners and mining companies the choice of selecting the most appropriate sizes. Cartridge length was not a critical issue for miners, the length ubiquitous for underground mining being eight-inches. However, explosives manufacturers offered six-inch cartridges for use in short holes such as those drilled by hand, or by coal miners who needed only a small charge. Lengths of ten and twelve inches were used in deep, machine-drilled holes requiring much dynamite. Long cartridges reduced the per unit time involved in loading explosives into drill-holes.

The cartridge diameter a miner chose was purely a function of drilling technology, and his main criterion was to match the diameter of the drill-hole. Small diameter cartridges, including three-fourths and seven-eighths inches, were most popular in mines where miners drilled by hand and holes were narrow. Mechanical rock-drills bored broad holes necessitating larger diameter cartridges, typified by one-and-one-eighth, one-and-one-quarter, and one-and-one-half inches.

Mines working with dynamite determined the correct quantity of cartridges to load into a drill-hole like those working with powder. The goal was to find a balance between minimizing the quantity of dynamite, and therefore cost, while maximizing the amount of rock blasted. Loading too little dynamite resulted in a partially or unevenly blasted working face, while loading too much dynamite fragmented the rock into unnecessarily small bits, and possibly compromised the integrity of the mine workings.

How did the miner conclude what the most efficient quantity of dynamite was? He loosely based his decision on the cumulative sum of a number of variable factors. These included the rock type, the size and depth of the drill-holes, the total area of the working face, whether any flaws existed in the working face, and the type of explosive used. Dense and hard rock required more dynamite than soft rock. Deep drill-holes or a broad working face necessitated a large quantity of dynamite. On the other hand, if there were relatively large flaws or weaknesses in the working face, a miner could use them to his advantage, allowing him to load less dynamite. Still, within these parameters there was considerable room for error and misjudgment.

Some mining and explosives engineers attempted to govern the amount of dynamite miners loaded into drill-holes with formulas. Although they meant well, engineers who subscribed to rigid quantification overestimated the practicality of their formulas and underestimated the ability of experienced, gritty miners. For example, an engineer based one formula on "units of strength," which he derived by multiplying the total pounds of dynamite by its percentage strength. For example 100 pounds of 40% dynamite converted into 4,000 "units." If those 100 pounds were loaded and shot and the face was completely blasted out, then the units balanced and the miner determined the correct quantity. If only part of the working face was blasted out, then more units were necessary[74]. The miner was supposed to figure increasing units by recalculating based on a higher percentage strength, or using more of the same. Another mining engineer suggested tracking the ratio of how many pounds a given percentage strength of dynamite were required to blast a square-foot of rock of a certain depth[75]. A superintendent, engineer, foreman, or even a miner could have assembled a chart based on the ratios of rock each percentage strength of dynamite blasted. Although constant figures could have been derived, this formula was of limited value because rock types in mines varied widely.

Similar to the above idea, engineers and miners devised a gross generalization of approximately how much dynamite blasted one cubic yard of a rock type. The formula claimed that one-and-one-quarter pounds of dynamite blasted a cubic yard of slate, one pound for sandstone, two pounds for conglomerate, three pounds for rhyolite, and four-and-one-half pounds for schist[76]. This was perhaps the most practical formula for miners and engineers because there was little calculation involved, and it was broadly applicable.

Some mining engineers expressed doubt regarding the practicality of formulas, and further they felt experienced miners were actually the best judges of how much explosives to load into a working face[77]. Although some miners used constant figures for rock types, in most cases their judgment steered them in a slightly simpler direction. In actual practice, hardrock miners tended to load their drill-holes approximately three-quarters full with dynamite, regardless of differences in rock types, and in some western mines where the rock was very hard, miners loaded drill-holes to within six inches of the collar[78].

Based on the knowledge of how much dynamite miners loaded into a drill-hole, the modern-day historian can determine the approximate quantity of dynamite consumed in a single blast of driving a tunnel or shaft. For example, miners drilled six-foot holes and they loaded four feet of each with dynamite. To do so, they had to pack in seven eight-inch cartridges. One cartridge of an ammonium nitrate/nitroglycerine product, such as dynamite extra, weighed as little as a half pound, while a cartridge of a dense explosive such as gelatin weighed as much as one pound. A cartridge of straight dynamite fell somewhere in between. Therefore, each drill-hole loaded with dynamite extra contained approximately three-and-one-half pounds of dynamite extra, or seven pounds of gelatin. If miners drilled the working face with a pattern consisting of four cut holes, eight trimmers, and four lifters, then one blast consumed approximately 65 pounds of dynamite extra or 112 pounds of gelatin, which translates to one-and-one-quarter and two-and-one-quarter boxes, respectively. The equivalent quantity of straight dynamite fell somewhere in between these two figures. During an average shift, this was how much dynamite passed through the hands of a drilling and blasting crew.

Priming the Round

To miners, dynamite was worthless without blasting caps to provide the small shock necessary for sure and controlled detonation. Miners referred to caps as *primers*, and one cartridge *primed* with a cap and fuse was sufficient to detonate the other cartridges loaded into the drill-hole. As part of their daily shifts in the noisy, stifling underground, miners had to make up primers, one for each drill-hole they had bored into the working face of a tunnel, shaft, or stope. Priming was fairly straightforward, and shortcuts and errors could lead to unforgiving results. Because priming cartridges was of such importance, explosives and mining engineers felt that they had to establish guidelines for safe procedures. In most cases these procedures required what many miners thought was an unnecessary amount of time, driving them to find quick and potentially dangerous ways of priming cartridges.

Miners used two different types of blasting caps for making up primers. The first was the standard cap which was detonated with safety fuse, and the other was the electric cap, which did not experience popularity until the 1910s. The specific methods miners used to prime cartridges differed for each type of cap. Priming dynamite cartridges with standard blasting caps and safety fuse was the oldest, least expensive, and most popular method among underground miners. They developed the process when dynamite made its appearance in the 1870s, they tuned it in the 1880s and 1890s, and some miners continue to use it today.

Miners primed cartridges in any one of a number of select locations at a mine. This activity almost always left telltale archaeological evidence r e m a i n i n g today in the form of blasting cap tins, bits of safety fuse, fuse and cartridge w r a p p e r s, dynamite boxes, and sawdust packing from the boxes. Miners favored a b a n d o n e d cross-cuts and niches near points of work to prime cartridges because such places were sheltered from traffic, and from being used for other purposes. In prospect shafts and short adits lacking niches, miners

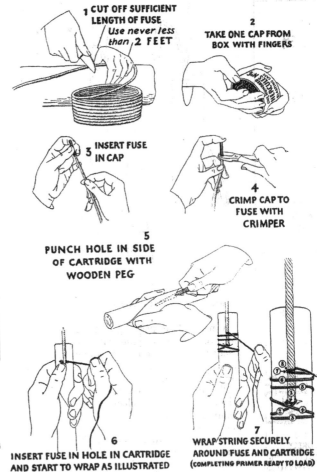

Figure 60. Priming a dynamite cartridge required several steps, including cutting an appropriate length of fuse, crimping a blasting cap to the fuse, piercing the dynamite cartridge, inserting the cap, and securing the fuse. Out of haste miners often sought to minimize these steps. The line drawing illustrates the side-prime method approved by engineers. (Courtesy Hercules Inc.; Source: Explosives Engineer Feb. 1931, p55).

primed their rounds on the surface of the waste rock dump, in a bunkhouse, in the hoist house, or in other mine buildings.

Once the miner had established a place to prime his cartridges, he began readying the tools and materials to complete the task. He first opened the tin of blasting caps, cracked open the box of dynamite, and unwrapped the coil of safety fuse. The miner also got out a cap crimping tool and a sharp jackknife.

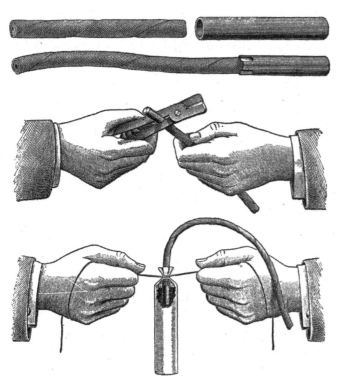

The *crimper*, also known as a *capper*, was the only tool mining and explosives engineers recommended for fixing a blasting cap onto the end of safety fuse when making a primer. Crimpers were similar in appearance to pliers, but when closed their jaw formed a circle slightly less in diameter than a blasting cap, so that it tightly crimped the cap's skirt onto the fuse. Crimpers came in a variety of styles - some included wire and fuse cutters on their heads, some were made of stamped sheet steel, while others were cast or machined. Most crimpers featured the name of the explosives company that supplied them. Although substituting other tools for crimpers was difficult because of their unique function, in many cases poorly financed miners used regular pliers and in some cases their teeth![79]

Figure 61. Early lithograph demonstrating the process of capping gutta percha safety fuse and end-priming a dynamite cartridge. The end-prime was probably the oldest priming method. (Source: International Correspondence Schools, 1907 A41, p87).

The miner continued to ass-emble primers by cutting appro-priate lengths of safety fuse with nice, neat, square cuts on a makeshift cutting board, such as a dynamite box lid. Western miners typically used six to eight feet lengths of fuse, which

Figure 62. Despite engineers' recommendations of proper end and side priming methods, miners favored inserting the cap into the cartridge's side and lacing the fuse back through, because it skipped a number of steps and saved time. (Source: Author).

allowed them plenty of time to light the entire round and calmly retreat. In the next step of priming, the miner slipped a blasting cap over the clean-cut end of one of the fuses, and crimped the cap's skirt with the crimping tool. If the fuse end did not have a neat cut, or if the crimp allowed the cap to wiggle loose, the miner had set himself up for a potentially deadly misfire. In wet conditions, mining and explosives engineers recommended sealing the joint between the cap and fuse with a special sealant or blasting soap. Too many miners neglected to cut fuse properly, crimp the cap adequately, or seal it against water, which caused the charge to misfire. Some miners paid with their lives and limbs trying to extract missed charges.

Last, the miner had to insert the blasting cap into the dynamite cartridge and secure the fuse. Here is where miners' actual practices diverged from the recommendations of mining and explosives engineers. To permanently anchor the blasting cap in the cartridge and to protect the protruding fuse from being damaged when a miner tamped the charge into the drill-hole, engineers recommended placing the cap in one of two places. For the *side prime*, a miner punched a small hole into the side of the cartridge, and he inserted the cap. The other recommended place for the cap was in the end of the cartridge, and it was known as the *end prime*. The crucial step in these methods was firmly lashing the fuse down with twine so the cap could not be accidentally pulled out of the cartridge. Of the two priming methods, many engineers favored the side prime because when miners tamped the charge down, which they usually did with great force, the fuse was less likely to be kinked and damaged.

In reality, miners, in a hurry to load and shoot their rounds and call it a day, did not like taking the time to fumble with twine and lash down fuses, especially when they had to prime fifteen to thirty cartridges. Instead, miners took a short cut that engineers unanimously recommended against. They stuck the cap in the side of the cartridge, but instead of tying the fuse down they laced it back through a second hole they punched through the cartridge's waste. The danger of this, claimed the engineers, was that the fuse could kink or split open, and set the dynamite on fire at least, or cause a misfire or hangfire at worst. But because this method saved over-worked miners precious time and could be directly attributed to few accidents, hardrock and coal miners across North America favored it[80].

Electric Blasting

From the 1880s to the 1910s, visionary mining engineers touted electric blasting as being superior to standard blasting caps and fuse for firing dynamite charges. Electric caps were relatively fool-proof because they came from the manufacturer with lead wires assembled and ready for use, they were waterproof, and the lead wires lent themselves well to being wrapped around a dynamite cartridge for securing the cap. Since they required no assembly on the part of miners, electric caps presented less risk of causing a misfire, and miners spent less time priming cartridges with them. Similar in size and shape to standard caps, electric caps featured two six- or eight-foot lead wires protruding from their ends, and when an electric current passed through the wires, it heated a hair-like platinum bridge nested in the cap's mercury fulminate charge, detonating it.

The popularity of electric blasting was practically non-existent in the first decade of dynamite's existence, and it grew at a snail's pace until after the turn of the century. The root of the dissatisfaction among mining companies with electric blasting lay in both the technology and the economics. Alfred Nobel developed the first electric blasting cap in the 1860s as a means of detonating dynamite, but he shelved the device for a time in favor of standard caps and fuse because it was too costly to make, and the generation of a reliable current was unsure. Notable explosives pioneers, such as H. Julius Smith who blasted on Massachusetts' Hoosac Tunnel railroad project; T.P. Schaffner who was president of the United States Blasting Oil Company; Jabez B. Dowse, Charles A. Browne, and George Mordey Mowbray who also blasted on the Hoosac tunnel, took up where Nobel left off. Mowbray is credited with developing the first truly functional electric blasting cap, which consisted of a mercury fulminate charge and electric bridge wrapped in gutta percha placed in a second charge, encased in a copper tube[81]. H. Julius Smith and Charles A. Browne were hot on Mowbray's heels with their own electric caps, and they signed contracts with the Laflin & Rand Powder Company and Oriental Powder Mills, respectively, two of the United States' leading explosives makers, to distribute their devices.

Hand-in-glove with the development of the electric cap, explosives engineers experimented with several sources of electrical current to detonate their caps. Engineers at first tried lead-acid batteries, but the huge glass vessels proved fragile and cumbersome to move in the rocky, confined workings of mines and railroad tunnels. They next turned to mechanical means to create a current, which could withstand the rigors of work underground. Most of the early development of generators occurred in Europe, but it was a North American inventor who adapted then-current technology into something usable for electric blasting. In the mid-1860s Moses Farmer built the first practical blasting machine in North America, which consisted of a hand-cranked dynamo encased in a wood box, which weighed 120 pounds[82]. While Moses' machine was able to set off electric caps in demonstration blasts, its large size and hefty weight left much to be desired.

H. Julius Smith, one of electric blasting's grandfathers, took the next step toward improving blasting machines, and he ultimately created a device which stayed with the mining industry for decades. Smith experimented with blasting caps and machines through the 1870s, patenting several versions of each. The blasting machine Smith settled on was a wood box measuring five inches wide, eight deep, and sixteen high that encased a small but powerful dynamo, which weighed less than thirty pounds. The dynamo was activated by rack and pinion gearing, and to use the machine, the operator simply drew the rack up by its wood handle and slammed it down. The rack bar turned the dynamo, which generated a current. Smith's machine would only be modified but never replaced for mining. A few blasting supply houses in the East, and the Giant Powder Company in the West, offered Smith's blasting machines through the 1880s. During the 1890s electric blasting equipment began to increase in popularity. Blasting machines of this era took one of three forms. The most popular machine was similar to Smith's design, and it featured two brass or bronze terminals on top, a leather carrying strap, a builder's name-plate, and the rack bar. The second version, less common, had similar features but it was activated by pulling up on the rack bar, rather than pushing it down. This supposedly was a safety feature. The third form of blasting machine manufactured between the 1880s and 1910s had a stout crank handle projecting from the side to turn the dynamo. This version was least popular. All three types of machines weighed less than thirty pounds, they were compact, and were able to withstand rough treatment from miners and quarrymen.

A less popular blasting machine released in the 1910s was an ovoid hand-held model activated by twisting a detachable handle. The machine's shortcoming was that it could only fire ten blasting caps at once, which was inadequate for hardrock mining, but well-suited for coal mining. The improved push-down and twist model blasting machines remained virtually unchanged from the 1920s until the 1960s.

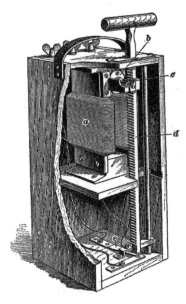

Figure 63. Cut-away view of a circa 1900 blasting machine. Note the three terminals on top; the outer two were positive and the center was a ground. When a miner pulled up and pushed down the handle and rack bar d, it engaged a pinion gear c which turned a rotor b. The rotor, rapidly spinning between magnetic coils a, generated an electric current. (Source: International Correspondence Schools, 1907 A36, p51).

Figure 64. Few miners or companies used special electric blasting kits because they required capital to purchase. Some large outfits did buy or make such kits, however. The illustrated kit includes a spool of blasting wire, a feed pulley, electrical terminals, and a supplies storage compartment. (Source: International Correspondence Schools, 1907, A36 p52).

From the 1880s into the 1910s blasting electrically was not popular among mining outfits for four major reasons. First, purchasing a blasting machine, electric caps, and other specialized supplies required capital, which many mining companies either did not have or would not spend. Why should they spend the money when in the short run standard caps and fuse cost least? Second, in remote mining districts electric blasting equipment was scarce, and miners who knew how to use it were fewer. Third, until approximately 1910 most available blasting machines were not suited for shooting the number of holes in the firing order required for hardrock mining. Most blasting machines predating 1910 had a ten cap capacity, while blasting most hard rock working faces required between ten and thirty charges[83]. As early as the 1880s thirty cap capacity machines were manufactured by companies such as H. Julius Smith and Laflin & Rand, but they were exceedingly rare and had to be special-ordered, discouraging would-be electric blasters. Fourth, a blasting machine detonated blasting caps simultaneously. The only way to adapt the machine to the firing order required for hardrock mining was to wire an expensive circuit delay box into the line, which was not widely available until the 1900s[84].

During the 1910s several events came together to make electric blasting efficient and easier to conduct, stimulating a rise in its popularity. First, shortly before 1910 H. Julius Smith and a few other explosives geniuses devised several varieties of delay-action electric caps. The caps had miniature fuses inside them which were ignited by the platinum bridge, delaying the cap's detonation by the length of the fuse. By 1910 leading explosives makers offered three different delays, at last granting hardrock miners their firing order. By the 1930s, manufacturers offered ten different delays. Beginning in the 1910s mining and explosives engineers, explosives makers, and the Institute of Explosives Makers pursued a campaign aimed at improving the safety of blasting. Some engineers felt electric blasting was one way of achieving this goal, and they set about educating miners and their bosses on the merits of electric blasting. Last, makers of blasting machines upgraded their basic machines to a thirty cap capacity, enough to shoot most working faces in hardrock mines. With the improved technology, an increased number of large underground mining operations experimented with blasting electrically, and the practice gained popularity through the 1920s.

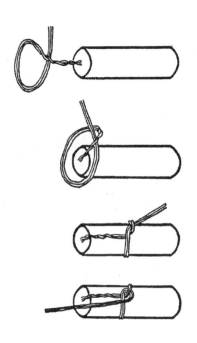

The process of priming dynamite cartridges with electric blasting caps was similar to priming with standard caps and safety fuse. The most popular method was the side-prime, and the cap's lead wires were tied in a half-hitch knot around the cartridge's waist. Another method that miners used involved inserting the cap into the end of the cartridge and passing the lead wires back to its middle and through a hole in the cartridge's center. A third method of priming combined the above two ways, but it was less popular because

Figure 65. The recommended method for priming a dynamite cartridge with an electric blasting cap. (Source: Author)

it took longer. Priming dynamite cartridges with electric caps was so straightforward and simple there were few if any shortcuts miners could take. Notably, these procedures comprised one of the few realms in mining where the methods espoused by mining and explosives engineers agreed with the methods actually used by miners.

Loading the Round of Charges

After making up primers and figuring out how much dynamite they needed, the drilling and blasting crew began loading the cartridges into the drill-holes. As with priming, the procedure for loading the drill-holes was straightforward, but in their haste to finish for the day and return to ground-surface, some miners had the propensity to hurry, which caused them to make mistakes that resulted in misfires and life-taking accidents.

The loading process began after the mining crew had drilled the blast-holes and removed their drilling tools from the heading, and fetched the boxes of dynamite and loading implements. The team of miners first neatly laid out the primed dynamite cartridges, which were equal to the number of drill-

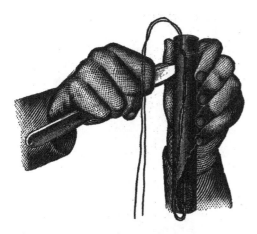

Figure 66. Except when working in watery drill-holes, miners had to slit dynamite cartridges end-to-end to facilitate expansion when they tamped the charges. Hence most miners carried sharp jackknives. (Source: Kirk, Aurthur, 1891, p23).

holes, and then they started plucking cartridges out of the box and slitting them lengthwise with jackknives so the cartridges could expand and fill the drill-holes when tamped[85]. Until the late 1930s, explosives manufacturers made dynamite cartridges with heavy waxed paper, which held its shape well and protected the contents from moisture. The cartridges were so rigid that miners had to slit them, which consumed time and gave the miners nitro headaches through accidental dermal contact with exuded nitroglycerine. In the late 1930s manufacturers began making cartridges under names such as *Tamptite, Redislit, EZ Load,* and *Perforated* with perforated wrappers designed to break open during tamping. Perforated wrappers eliminated the unpleasant task of slitting cartridges. In the next step of loading dynamite, one of the miners began inserting slit cartridges and one of the primers into a drill-hole, and the other miner used a tamping rod to firmly pack them in, leaving no air space that might lessen the explosion's effect. Last, one of the miners sealed the remainder of the drill-hole with stemming material in the same manner as they did when loading blasting powder charges.

Tamping cartridges was not without risks, and in the first two decades of using dynamite in hardrock mines, miners climbed a steep learning curve of death and carnage. All too frequently miners went against common sense and recommendations of mining and explosives engineers and used steel tamping rods or drill-steels instead of wooden rods, often with violent, pounding action. Tom Watson, who worked in a California gold mine with a father-son drilling and blasting team in the 1930s, experienced firsthand what befell many miners decades before him as he watched the boy dying in the arms of his father. The boy made the mistake of using a six-foot drill-steel to ram home fairly old nitroglycerine-based dynamite. The energy he was expending upon the nitroglycerine was enough to detonate it, which shot the drill-steel through his abdomen[86]. This boy was not alone, many before him did likewise. Tamping dynamite with steel rods had been known to be the root of many accidents, prompting the state of Ohio to pass a law in 1907 forbidding the practice. Even when abiding by engineers' recommendations, tamping was dangerous. Miner John Gurcham was using a wooden tamping

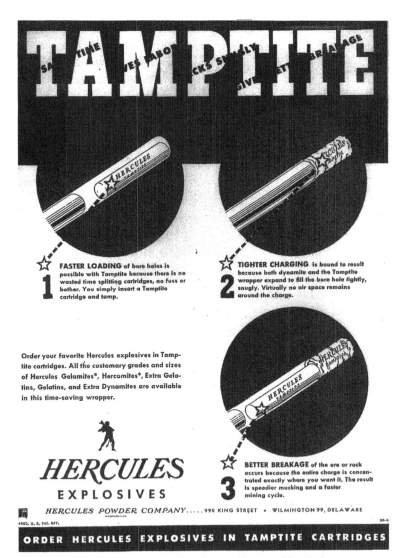

Figure 67. In the late 1930s the Hercules Powder Co. developed a perforated waxpaper dynamite wrapper designed to break open when tamped into a drill-hole. Many miners liked the wrappers because they no longer had to slit cartridges, and mining companies appreciated the savings in time. (Courtesy Hercules Inc.; Source: Explosives Engineer *Jan-Feb. 1944, p47).*

rod in 1926 when a fully loaded drill-hole exploded, the concussion and shattered rock killing him[87]. Gurcham used excessive force to tamp dynamite known to have been at least four years old, and it was just too sensitive for such force.

There was no hard consensus between miners and mining engineers regarding the order in which to load the primed cartridge among the column of non-primed cartridges in a drill-hole. In agreement with the

recommendations of some mining and explosives engineers, most miners loaded the primer last because this position was most straightforward and safest. Loading the primer first increased the risk of damaging the fuse because it lay exposed the length of the drill-hole, which was a problem especially when a miner was tamping charges and stemming material vigorously. And, as stated earlier, damaged fuse was the greatest cause of the miner's dreaded misfire.

From a technical standpoint, loading the primer last resulted in a more efficient blast, because the explosion started at the top of the hole and propagated down the rest of the charge, breaking the outer rock first, releasing the pressure for the cartridges tamped deep in the drill-hole. Some miners felt that the primer

Figure 68. One miner inserts electrically primed dynamite cartridges into drill-holes in a tunnel heading while a second miner tamps them in home under Puget Sound in 1958. The tamping rod appears to be a steel pipe probably tipped with a wood plug. (Source: Hagley Museum & Library).

was best loaded in the center of the charge for a symmetric blast, and other miners felt that placing it first, at the bottom of the drill-hole, was best because the explosion fractured the bottom rock first, and the rest of the rock was shattered as the explosion moved up and out[88].

When working in very hard rock, such as granite and metamorphic material, some miners attempted to

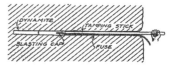

Figure 69. Side view of a drill-hole properly loaded with dynamite prior to being sealed with stemming material. Note how very aggressive action with the tamping stick has the potential to damage the fuse and dislodge the blasting cap from the dynamite cartridge. (Source: E.I. DuPont de Nemours & Co., 1920, p35).

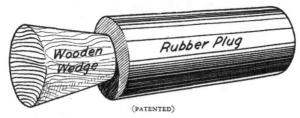

(PATENTED)

The Heitzman

for: **Safety Blasting Plug**

Safety:

1. Misfired holes are handled without withdrawing the charge or drilling an extra hole. A second primer is placed in the open hole of the unexploded charge, sealed with a second plug and fired. The whole charge explodes.

2. Does not injure the wires or fuse in expanding against the walls of the drill hole.

Efficiency:

1. Provides a uniform stemming.

2. No dirt stemming is necessary.

3. Seals the hole airtight.

4. In pitch work, the charge is easily sealed.

Economy:

1. Produces large lump coal, and a great percentage of prepared sizes.

2. Produces maximum yield of coal, ore, or rock with no increase in explosives.

3. Saves time and cost making "dirt bags" for tamping.

4. Saves the time and work to drill extra holes in case of misfires.

5. Saves the extra explosives needed when a shot misfires.

The *Cost* of the Plug does not equal the *Advantage* in *Efficiency* and *Safety* it Produces.

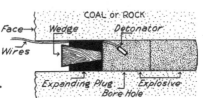

When ordering give the size of the drill hole.

Heitzman Safety Blasting Plug Co.
1310 West Pine Street   Shamokin, Penna.

Figure 70. For a brief time during the mid-1930s, mine supply houses offered the Heitzman Safety Blasting Plug as a superior alternative to the clay and drill cutting stemming materials miners traditionally used. Because the hole plug cost money and it did not work well with safety fuse, it never saw popularity. The device consisted of a rubber cup which a miner inserted into a drill-hole after loading the explosive charge, and wood plug that caused the rubber to expand, when hammered into the cup. (Courtesy Hercules Inc.; Source: Explosives Engineer *Dec. 1935).*

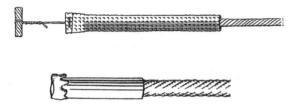

Figure 71. During the 1900s and 1910s entrepreneurs developed several devices to assist the miner in igniting safety fuse. The first was a match head in the form of cap that slipped over the end of a fuse. A miner needed one cap per fuse end. The other device was the wire pull igniter, which a miner could have used to ignite several fuses. (Source: E.I. DuPont de Nemours & Co., 1920, p41).

mix dynamite percentage strengths in drill-holes in hopes of maximizing efficiency while minimizing costs. One successful practice miners worked out consisted of loading several cartridges of 60% dynamite at the bottom of the hole, and cartridges of less-costly 40% in the remainder. The idea was that the 60% dynamite would shatter and crack loose the bottom rock while the 40% dynamite was adequate enough to break the top rock and move it out[89].

Low-budget operations, especially during the Great Depression, exercised a special but dangerous loading practice in hopes of saving money. Drilling and blasting crews staggered ceramic or wood spacers, known as *dummy cartridges*, between live cartridges to reduce the quantity of dynamite loaded into a drill-hole. What happened all too often was that the explosion did not propagate to all cartridges in the drill-hole, resulting in unexploded dynamite mixed in with the muck. Most mining companies had the sense to provide their miners with enough dynamite to discourage this dangerous practice[90].

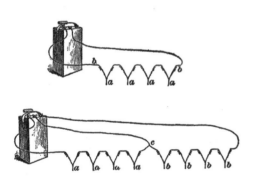

Figure 72. For electric tunnel and shaft blasting with a conventional quantity of charges, miners wired the electric caps to the blasting machine with a series circuit. For large shots prior to the 1910s, miners used a special blasting machine featuring two positive terminals and one common ground terminal for wiring two separate series circuits. After the 1910s miners used parallel circuit wiring, instead. (Source: International Correspondence Schools, 1907 A41, p93).

Fire in the Hole

The procedure that miners used to shoot a round of holes loaded with dynamite primed with standard caps was unconsciously carried over from the era of blasting powder. To shoot a round of dynamite charges, a miner made a rat tail from a small length of fuse and he lit each hole in the firing

Figure 73. Soggy miners under Puget Sound are connecting electric blasting cap wires in preparation to fire a round of charges in a tunnel heading under Puget Sound in 1958. They appear to be preparing a series circuit. (Source: Hagley Museum & Library)

order. Early in the 1910s explosives manufacturers introduced the *wire-pull igniter* as a substitute for rat tails[91]. This ingenious device was a small fiberboard tube of slow-burning powder which the miner slipped over the end of a safety fuse to light it, and the igniter burned long enough to enable a miner to light a number of fuses. Although superior to the rat tail and the flame of carbide lamps for lighting fuses, the wire-pull igniter was an accessory which cost money, while the rat tail was virtually free, and as a result most mining companies leaned toward the less-costly rat tails. Generally only well-capitalized mining companies and large blasting projects purchased wire-pull igniters[92].

To simplify the process of lighting numerous fuses at once, some well-financed mines began using Thermalite Connectors as early as the 1940s to link the plethora of fuses necessary to shoot large blasts. Miners crimped the connectors to the ends of safety fuse in the same manner as standard blasting caps, and the connector's head was bent over a notch cut into a second piece of fuse to form a "T". In so doing, numerous fuses could be linked to several trunk lines.

Shooting a round with electric caps was a relatively new process developed for blasting with dynamite, and not an adaptation from the days of using

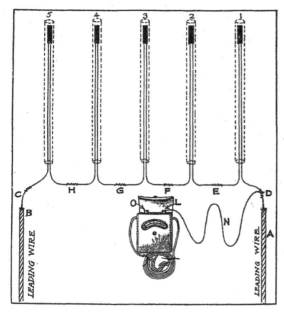

Figure 74. Mining companies with an interest in safety and abundant capital supplied their blasting crews with galvanometers for testing electric blasting circuits prior to firing. The diagram illustrates how to test a blasting circuit for defects with a galvanometer. A miner tested each blasting cap by grounding the device through wire N to one of the trunk wires, and touched the positive wire to the junction of each cap, represented by C, and E-H. (Source: E.I. DuPont de Nemours & Co., 1920, p71).

blasting powder. What a miner faced as he was looking at a tunnel heading loaded and primed with electric caps was an array of sealed drill-holes with two wires hanging out of each. To prepare the shot, the miner simply connected the lead wires together in a series circuit, and if the entire round contained a huge number of charges, he wired them together in a parallel circuit. Once the miner checked and polished all of the connections, he connected the two remaining free wires to spools of blasting wire which he played out to the nearest point of safety.

Figure 75. Line drawing of a typical galvanometer. (Source: E.I. DuPont de Nemours & Co., 1920, p69).

If the miner's company had the luxury of buying proper blasting supplies, then he may have used a galvanometer to check the resistance of the blasting circuit. A high reading on the device's meter meant that there was resistance, which constituted a problem with the wiring. Once satisfied, the miner connected the free wires to the terminals on the blasting machine, shouted "fire in the hole!" and slammed down the handle. The machine sent out an electrical charge heating the platinum wires in the caps, which exploded, detonating the dynamite. The drilling and blasting crew listened expectantly for the reports of the shots, and when satisfied that the charges fired, they grabbed their lunch buckets, went to the surface, and greeted the next shift beginning their day of one round in and one round out.

CHAPTER THREE

MINES, MINERS, AND EXPLOSIVES

Unquestionably, blasting powder and dynamite revolutionized the practices of mining; both types of explosives offered benefits well beyond any tool the minerals industry had yet seen. In addition to the positive implications such a profound technology held for the minerals industry, it had a direct and personal impact on miners by dramatically changing their work environment. This chapter discusses the four most significant issues explosives posed to miners and mining companies, which were storage, transportation, misfires, and gases. Here miners' actual practices are compared with the recommendations of mining and explosives engineers. The above issues shared relationships and affected one another. For example, improper storage of dynamite degraded or froze it, which led to misfires and noxious gases.

August 16, 1900, began as an ordinary day in the Cripple Creek Mining District. Steam hoists chuffed, boilers hissed, and the ring of blacksmiths' hammers could be heard coming from many of the mines. The blacksmith at the Eclipse No.2 shaft on Raven Hill was busy sharpening drill-steels for the miners when he accidentally knocked coals out of his forge. As was common practice among Western hardrock mines, the forge, hoist, boiler, and equipment storage were all enclosed in the shaft house. So was the mine's supply of dynamite—several dozen boxes—and unfortunately the blacksmith had knocked coals onto them. Rather than heroically try to extinguish the smoldering embers, he sounded the alarm and fled. Within a short time the embers started a fire which ignited the dynamite, and it exploded, blowing the shaft house across the hillside, and wrecking other parts of the mine plant[1].

Any mining operation that used explosives had to store them some place. Well-financed mining companies and construction contractors often stored their dynamite and blasting powder in proper explosives magazines, but many did not. What most Gilded Age mining companies considered adequate storage often contrasted sharply with what explosives and mining engineers considered proper, and the event at the Eclipse No.2 shaft illustrates what could happen.

> The storage of explosives has a much deeper relation to safety in their use than is commonly realized. Improper storage of explosives, detonators, fuse and squibs leads directly to misfires and to the incomplete detonation which leaves unexploded dynamite in the bore-hole or thrown out among the blasted material, and to the burning of charges in the bore-hole[2].

As can be inferred from the above quote, explosives storage directly affected the lives of miners. Improper storage degraded all explosives and permitted dynamite to freeze, hampering the performances of the explosives and causing problems such as misfires, partial detonation, and poisonous and noxious gases.

In the eyes of explosives and mining engineers, explosives magazines had to fulfill a number of specific requirements, and persons who entered magazines had to exercise the utmost caution and judgement. While large mining companies could afford to build special buildings and spare money and manpower maintaining them, smaller outfits were much more lax about how they stored their explosives—either through ignorance, a laissez faire attitude, or economic necessity.

Proper magazines came in a variety of shapes and sizes; from small armored boxes on rollers and portable structures, to concrete blockhouses and large earth-covered bunkers. Explosives engineers felt that all magazines must be bullet-proof, fire-proof, dry, and well ventilated. They felt that magazines should be constructed of brick or concrete and, if of frame construction, be sand-filled and sheathed with iron[3]. These features not only protected the explosives from physical threats, but they also regulated the internal environment, which was important, especially in summer. Extreme temperature fluctuations were proven to damage explosives, and moisture spoiled ammonium nitrate, extra formulas, blasting powder, railroad powder, and straight dynamite. Explosives engineers also felt that a magazine's location was best situated in an open, cleared area away from other structures, railroads, and highways.

Regardless of the recommendations of engineers, mining companies throughout North America stored their explosives in very crude and even dangerous facilities. Many mines used sheds sided only with corrugated sheet-steel where protection from fluctuations in temperature were minimal. Magazines such as these allowed dynamite to freeze in the winter, and in the summer they became hot-houses where ammonium nitrate desensitized and nitroglycerine separated out. Worse yet small, capital-poor operations stored their explosives in dug-outs merely roofed with sheet-steel, or in abandoned prospect adits. Jared Stark, field representative for DuPont, documented a particularly hazardous magazine. He found one company in West Virginia's Pocahontas Coal Fields using a wooden shed located only yards from the mine's headframe and not much farther from a main rail line[4]. Not only was the magazine made of flammable building materials, but it was proximal to sources of cinders from the mine's boilers and passing railroad engines. If it exploded, it would have caused great damage and probably loss of life.

Storing explosives underground in a mine was a different situation. Although probably the single most popular practice, it was not heavily addressed by mining and explosives engineers except to say mines tended to be damp, a factor that affected explosives over time. Despite dampness, at least in the West, most mining companies dedicated an abandoned drift or two as magazines. In fact, Utah gold miner Ray Sporr worked for a mine near Salt Lake City that deliberately drilled and blasted two rooms off its main tunnel for magazines[5]. Many large mines had a mother magazine on the surface which received explosives shipments and distributed batches to satellite magazines underground. Storing explosives underground was beneficial because not only was the temperature constant through the seasons, but in the West most mines tended to be dry.

For some miners storage was even less formal than corrugated steel sheds and prospect adits. Contract miners and crews working for small development companies all too frequently stored up to hundreds of pounds of dynamite in their cabins and bunkhouses. The manager of one Colorado mine thought he could kill two birds with one stone and had the mine's dynamite stored in a place where it would stay dry and thawed. He chose the mine's steam boiler and had ten boxes placed underneath and two on its back, where they became too hot to touch[6]! With such crazy storage it is a wonder that there were not a great many accidents. Fate turned a blind eye toward these operations, which is why such foolhardiness continued with little correction.

Miners' behavior in the magazine was also important regarding safety because careless actions and negligence could cause at worst a cataclysm, or merely damage the explosives to such a degree that they misfired or incompletely detonated, presenting a hazard to other miners. Explosives engineers emphasized that miners not smoke and that all tools inside the magazine be of non-sparking materials. More importantly, they emphasized keeping the floor swept of loose powder, scrubbed of leaking nitroglycerine, and order overall. Engineers unanimously agreed that magazines should be used for storage only and that no kegs or boxes should be opened therein, and under no circumstances should cartridges be made-up or primed. They equally stressed that caps and fuse needed to be stored in a totally separate magazine.

Figure 1. Beginning in the 1890s mine supply houses began offering prefabricated explosives magazines. Consumers of prefabricated magazines, such as the structure illustrated, tended to be small companies with capital allowing for such expenditures. Most mining companies, however, had limited funds and skimped on storage facilities. (Source: Kirk, Aurthur, 1891, p141.)

As was par for the course, many miners did not adhere to these rules. Jared Stark saw numerous magazines throughout Pennsylvania with loose powder and nitroglycerine stains on their floors and thirty to forty kegs of powder with their bungs left off for the miners' convenience[7]. Mining engineer Walter Johnson and gold miners Ray Sporr and Tom Watson remembered when boxes of dynamite were opened and left in magazines. In addition, it seemed the rule rather than the exception throughout North America to have stored caps, fuse, and open dynamite boxes together in the same magazine[8]. Mines practiced this behavior from the 1870s into the 1940s.

Management in many small mines in the West felt that intense supervision of magazines and miners' behavior in them was unnecessary. They assumed most miners were reasonably careful and responsible. But in larger mines where dozens of miners, at the least, came and went, some of whom tended to be irresponsible either through laziness or incompetence, control was necessary. In response, mining companies created the positions of magazine-keeper and powder monkey to serve as a means of regulation. A magazine keeper differed from a powder monkey in that the keeper merely attended to the magazine while the powder monkey did likewise with the added responsibility of blasting for the miners. Well-built magazines were a hallmark of well-capitalized, ore producing mines, while mines with limited financing tended to have shoddy magazines or none at all. Likewise, only large, well-capitalized mines could afford magazine keepers and powder monkeys. These positions started to become more common around the turn of the century. By hiring powder monkeys and magazine keepers, mining companies felt they had a better handle on storage, transport, and use of explosives. The result, they hoped, was monetary savings in time and explosives, a reduction in poisonous byproduct gases through more efficient use of explosives, and improved overall safety. Still, with the plethora of pay-as-you-go mines, miners continued to provide their own storage, proper or inadequate, they transported explosives safely or otherwise, and conducted their own blasting, efficient or not, into the 1950s.

Figure 2. The Blue Flag Mining Company at Cripple Creek, Colorado, was a large, well-financed organization, and it built a stout magazine to store dynamite. Masons constructed heavy mortared rock walls, an arched brick ceiling, and a second wood roof, all stabilizing the interior climate. (Source: Author.)

By the 1890s mining engineers grew increasingly aware of the hazards that mishandling explosives posed to the function of a mine, as well as to life and limb. As part of this awareness, they acknowledged the dangers that lurked in transporting explosives from the magazine, through a mine's busy support area on the surface, down the rough and jostling ride underground, through tight quarters filled with miners, to the working face. Engineers espoused what seemed to be obvious safety precautions, such as not smoking while carrying explosives, not

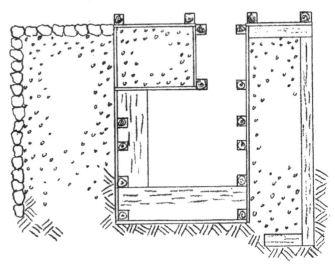

Figure 3. Plan view of the construction and materials of a well-built magazine. The structure consists of heavy, gravel-filled walls enclosing a chamber excavated out of a hillside, and roofed with earth. The magazine is located at the Enterprise Mine, Gold Mountain, Nevada. (Source: Author.)

Figure 4. Many large mining companies stored explosives in magazines in the underground workings, which were supplied from mother magazines above ground. A magazine keeper poses in front of stacks of DuPont Red Cross Gelatin boxes at the 2400-foot level in the Granite Mountain Mine, Butte, Montana, in 1920. Gelatin dynamite was a wise choice of explosive for work in Butte's wet mines. (Source: World Museum of Mining Photoarchives.)

carelessly tossing about boxes of dynamite, not opening kegs of powder near open flames, and only the person responsible for a load of explosives accompanying it down into the mine[9]. Yet, miners disregarded each and every recommendation, tragically in some cases. Their propensity to disregard dangers prompted an editor of DuPont's *Blasters' Handbook* to state: "In transporting or handling explosives, care should always be exercised. Do not let familiarity breed contempt"[10].

Generally, large mines with great income could afford the manpower and resources to regulate the transport of explosives. It was the smaller mines and contract miners that presented the greatest danger, usually to themselves. Some anecdotes are comical —almost. In 1923 a coal miner in Schuylkill County, Pennsylvania, was casually trans- porting dynamite cartridges in his shirt pockets to the breast where he was working. While underground he stumbled and fell face- down and his crashing weight detonated the dynamite[11]. Another miner in Lackawanna County, Pennsylvania was foolish enough to carry primed cartridges in his shirt pockets, as well. He loaded the drill-holes, stuffed the extra cartridges into his shirt pockets, spit the fuses, and retired to safety. He suddenly realized something was dreadfully wrong and

Figure 5. Railing against common sense and against the advice of mining engineers, in the early 1920s a Cleveland- Cliffs Iron Company miner begins to ascend a raise with a clutch of drill-steels in one hand and a bundle of dynamite cartridges bound with safety fuse slung over his shoulder. The miner also has a coil of safety fuse wrapped around his upper arm, and probably a tin of blasting caps in his pocket. This image exemplifies the "devil may care" attitude common among miners underground. (Courtesy Hercules Inc.; Source: Explosives Engineer *April 1924, p 125.)*

looked down to see too late that he accidentally lit the fuses to the cartridges in his shirt[12]!

Despite occasional accidents, miners learned to transport blasting powder with reasonable safety at an early date. By the 1860s miners blasting coal and

Regular Round   Round Comb. Squib Box   Round with Bail   Oval with 4 Lugs

Figure 6. Between the 1880s and 1920s mine supply houses offered four basic types of flasks for transporting small quantities of powder in mines. The left model, a style dating back to the 1860s, had a handle on the side for pouring out powder. The flask second from the left had a handle and a built-in squib tin. The flask second from the right was round in shape, had a handle on the side for pouring powder, and it had a carrying bail. The flask on the far right featured side-lugs for a shoulder strap, and was ovoid in shape. (The Martin Hardsocg Company, 1912, p 65.)

soft ore, rarely using an entire keg in a shift, transported small quantities of blasting powder in special flasks. Powder flasks were usually iron cans ranging from a one quart to one gallon capacity. Flasks were made in four basic varieties, and all of the metal types featured conical spouts covered with slip-caps. The rarest type was made of either pressed paper pulp, papermache, or pressed cork, and it featured a wood pouring handle pinned to the body, and a cork stopper. Its construction was intended to be non-sparking for safety. The second variety consisted of a steel cylindrical body and pour spout with a pouring handle soldered on. Some versions also had wire bail handles for carrying. The third type of flask was similar to variety noted above, except it had a built-in squib container for convenience. The last variety of powder flask had an ovoid body and canvas shoulder straps. Of all the types surviving today, cylindrical flasks are most common, partly because they were easiest for a mine blacksmith or tinsmith to duplicate, often using old food cans. As with squib tins, most blacksmiths and tinsmiths made powder flasks with soldered construction until around 1910, while some began to make flasks with rolled and crimped seams as early as the 1890s.

One inherent problem with the use of flasks was filling them. Their mouths were narrow and spilling powder was too easy, unless a person used a funnel. One old coal miner in Pennsylvania broke several rules and ended up learning a hard lesson. He was filling a powder flask from a keg in his house, and since he had smoked his pipe while doing this before, he saw no reason to do otherwise. He must have been pouring the dregs from the bottom of the keg because just enough powder dust rose in the air to catch fire and in the blink of an eye its flash ignited the remainder in the keg and flask. The resulting explosion blew the side of his house out, scarred his face, and burned off an ear[13]. Most accidents regarding the handling of powder are attributable to stupidity and miners' numbness to the omnipresent danger that explosives presented.

In hardrock mines where miners often used over twenty-five pounds of powder in a given shot, they typically transported entire kegs to the workings. Blasting powder came packaged in discrete quantities measured by weight, twenty-five pounds being the uniform standard. Although the twenty-five pound keg was most common, divisions of twenty-five including twelve-and-one-half and six-and-one-quarter pound sizes, and multiples of twenty-five, including fifty pound and one-hundred pound kegs saw use. Kegs which were twenty-five pounds and smaller were consumed by prospectors, contract miners, loggers, and land developers, who bought only enough powder to meet immediate requirements. Kegs of twenty-five, fifty, and one-hundred pounds were bought by outfits that consistently consumed high volumes of blasting powder, such as large mining and quarrying companies[14].

Dynamite had different handling requirements than blasting powder, and during the 1870s and 1880s miners had to learn by trial and error just what these were. Sensitivity to shock was one of dynamite's most notorious qualities which affected handling and transport. In the beginning miners had no idea that tragedy hovered over poor practices in handling dynamite. How could they when manufacturers and engineers claimed "it can be tumbled about as safely as so many boxes of soap," dynamite can be "hammered between wood and stone, wood and iron, and sustain any shock," and that it may be conveniently carried in boots and pockets[15]. In 1926 John Gurcham learned this lesson when he was tamping a charge of four-year old dynamite just a little too vigorously with a wood tamping rod, as recommended by engineers. The dynamite detonated because it was too sensitive and it killed him[16]. One experienced explosives engineer aptly stated: "Dynamite will stand treatment at one time which at another time will result in explosion," summing up dynamite's capricious nature in relationship to handling[17].

To alleviate the dangers of transporting dynamite cartridges, by the 1910s some large mining companies began supplying miners with special carrying cases. Some were shoulder bags made of canvas, rubberized canvas, or fiberboard capable of holding up to twenty-five pounds of dynamite. Powder makers sold knapsacks with 100-pound capacity, and the Hammond Safety Explosives Box, made in Pennsylvania, held several pounds, enough dynamite to shoot a coal breast. Carrying cases were no different from any other form of safety equipment or specialty tool in that their main market was mining companies and contract miners which had a conservative view toward safety and enough money to do something about it. Again, such companies tended to be well financed and productive.

In most mines, special carrying cases were an exception rather than the rule. In the West miners who needed a relatively small amount of dynamite often carried it in simple burlap sacks, and when they needed more, they

DUPONT

The Complete Line of
Powder Bags
for carrying explosives

E. I. du Pont de Nemours & Co., Inc., Newburgh, N.Y.
Branch Offices: Denver, Colo.; Duluth, Minn.

Figure 7. Like other large explosives makers, E.I. DuPont de Nemours & Company offered factory-made canvas bags capable of carrying up to fifty pounds of dynamite. (Courtesy E.I. DuPont de Nemours & Co.)

transported the explosive in the original wood packing boxes, either full or partially empty, to their workings[18].

The means by which miners conveyed explosives underground presented issues. In mines worked by lessees, especially large coal mines where haulage was often supplied on contract by a trammer or mule skinner, ore cars were not always available and miners had to hand-carry their boxes and kegs. On the shoulder seemed to be a favored way, despite the warnings of mining and explosives engineers. With dynamite there was the danger of dropping it and setting it off. If a keg of powder was dropped there was the danger that it could rupture and be ignited by a spark. Carrying kegs of blasting powder on the shoulder posed one additional danger often overlooked by miners. Twice in 1907 DuPont field representative Jared Stark documented accidents in which miners carrying kegs on their shoulders brushed up against overhead trolley lines for the mine's electric locomotives. If the high-voltage shock did not kill the miners immediately, the following explosions did[19]. Unfortunately seven other miners were also killed each time.

Some large mines supplied specialized cars that were basically armored, insulated boxes on rail trucks for transporting explosives from surface to underground magazines or to workings for big blasts. Such cars date back at least as far as the 1870s, varieties being used in Virginia City, Nevada, but they did not become common until after approximately 1920. Because such cars required time and money to construct, and because they represented handling of large volumes of explosives, such pieces of safety equipment were hallmarks of large, well capitalized, productive mines. Capital-poor mining companies would not have spent the time or money on special cars, or they did not handle enough explosives to justify the expense.

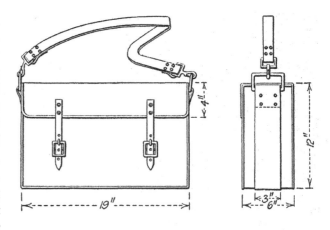

Figure 8. Plans for a fifty pound-capacity canvas shoulder bag for carrying dynamite through a mine. (Source: Engineering and Mining Journal *V.99, No.13, p 574.)*

The final step in transporting and handling explosives prior to loading them into drill-holes lay in opening their containers. Mining engineers suggested guidelines that appealed to common sense, such as opening only as many containers as would be used, not opening powder kegs in the presence of flames, not smoking, and gentle handling. Explosives engineers suggested that dynamite boxes be opened by prying their lids off with a hardwood wedge and wooden mallet. Of course, miners did not follow these recommendations and often smashed in box lids or sides, or pried them open with picks or like objects. In some cases, when opening dynamite boxes clever miners pried a two-inch slat off the top and tipped the box on its side so it could serve as a cartridge dispenser.

Figure 9. Mining engineers recommended carrying blasting caps in special safety carrying cases, which miners rarely used. The illustrated case can hold ten caps, which are accessed through a rotating dispenser top. (Source: E.I. DuPont de Nemours & Co., *1932, p 45.)*

Except for very violent action, most methods miners used to open dynamite boxes were not as much of a problem as with steel powder kegs. Usually, miners opened the kegs equipped with threaded bungs properly so they could be refunded for deposit, but with the thinner cleat-top kegs this was not the case. Miners, out of haste or laziness, took quick but potentially deadly means for opening cleat top kegs, as described by DuPont field representative Jared Stark: "I have no doubt, however, that many of our salesmen have had miners laugh

Figure 10. Well-financed, large mines employed powder cars, either purchased or custom-built, for transporting batches of explosives underground to magazines or for large blasts. The powder car illustrated was custom-built of wood by a coal company, using the chassis of a standard coal car. A miner is depositing a box of Hercules permissible dynamite into the car. (Courtesy Hercules Inc.; Source: Explosives Engineer *Aug. 1923, p 156.*)

at them, as one did at me in the New River coal field, in West Virginia, when I asked him to drop his pick long enough for me to reach a place of safety while he was tempting providence to shorten the days of his life by going through with such a foolhardy operation. I turned to walk away but had taken only a step or two when I heard the pick go through the keg top. What I said to that miner would not look well in print but no doubt would be approved by all present"[20]. Puncturing an iron keg with a steel implement was extremely risky because of the high possibility of throwing a spark. However, miners chronically opened kegs with this method because they felt it saved time and made a large hole for fast pouring. Why the written record is not full of accounts of miners blowing themselves to shreds by opening cleat top kegs with picks is a mystery.

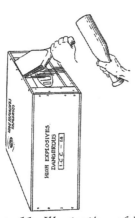

Figure 11. Illustration of how explosives engineers suggested boxes be opened with hardwood chisel and mallet. Miners rarely followed this practice and used instead whatever tools were at hand. (Source: E.I. DuPont de Nemours & Co., 1932, p 33.)

The morning of May 26, 1900, was a time which a hardrock miner named Davis never forgot. He and his partners reported for work at the Trade Dollar Mine in the Coeur d'Alene region of Idaho, and the three miners were warned of two misfires, charges of dynamite that failed to go off, in their tunnel heading. They planned to drill around the misfired holes, blast out the rock surrounding the charges, and pick them out of the muck. Davis and an assistant set up a rock drill and began drilling into the center of the face when the drill-steel hammered an unreported misfire. The charge promptly went off, seriously injuring Davis and killing his fellow miners[21]. Davis's misfortune was not an uncommon accident an America's mines, which is exactly why miners feared misfires.

When the drilling and blasting crew fired their rounds at the end of day, it was traditional for them to listen carefully and monitor whether all of the charges detonated. If they noticed a misfire, they usually passed word on to the next shift or the shift boss who investigated it after the heading cleared of gases.

Figure 12. This dramatic advertisement, featured in a 1944 edition of the Explosives Engineer, illustrates the collaberation between the explosives industry, the Institute of Makers of Explosives, and the mining industry to eliminate misfires. (Courtesy Hercules Inc.; Source: Explosives Engineer Jan.-Feb. 1944, p39.)

Here, observation of exactly which charge had failed and communication to the oncoming shift were everything.

The shifts that worked in the Gold Hill Mine in Quartzburg, Idaho, had an excellent method of communicating to one another about misfires. In each level of the mine there was a map at the shaft station showing the drifts and their working faces. Each working face featured three holes representing the cut round, the trimmer set, and lifter holes. If a given shift noted a misfire when they shot their rounds, they placed a peg in the appropriate hole in the map-board for the next shift to see[22]. Around 1910 the Monarch-Pittsburgh shaft in Tonopah, Nevada, instituted a form showing a schematic of the drill-hole pattern the mine used while sinking the shaft. The drilling and blasting crews coming off shift circled the holes they thought misfired and signed the

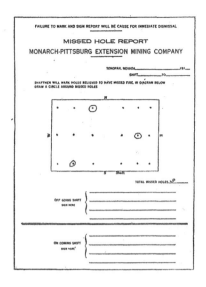

Figure 13. Several mining engineers at large operations devised special paper forms with which one drilling and blasting crew could communicate to the next shift regarding misfired shots. The form illustrated shows the wedge-cut hole pattern used when the Monarch-Pittsburgh Extention Mining Company was sinking a shaft in Tonopah, Nevada. The crew going off shift circled misfired holes for easy identification. (Source: Engineering and Mining Journal, 1916, p 56.)

form, while the crew coming on signed to acknowledge having read the form[23]. However, these attempts at communicating the occurrence of a misfire were far from common. Frequently a drilling and blasting crew left word of a misfire with the shift boss, who might fail to pass the information on to the next shift. Or they scrawled the missed hole on a blackboard, which few miners read and was often erased.

In most cases misfires physically manifested themselves in working faces as convex masses of fractured but unblasted rock. A little probing with a candlestick or the pointed end of a drilling spoon and a miner could locate the soft stemming material, such as drill-cuttings or clay used to seal the hole, and figure out exactly where the misfire lay. But this was not always the case as miner Davis's experience illustrates.

Misfires resulted from a variety of causes. With dynamite, a misfire could have been caused by a faulty blasting cap, a poor crimp on the cap might have allowed the fuse to separate, or water could have seeped in and spoiled both cap and fuse. Usually, though, the problem was kinked or damaged fuse, a cap dislodged during loading, or miners trying to shoot frozen dynamite[24].

Kinked and damaged fuse was also often the root of a less common but no less terrifying problem miners termed *hangfire*. A fuse hungfired when its internal flame stalled and temporarily, delicately smoldered until the powder train once again engaged. Titus Hibbert, superintendent of the day shift at the Jackson Iron Mine in Michigan, and one of his level foremen were walking back to a working face they tried to shoot to find out why, after so much time had passed, the round failed to detonate. As they turned a corner into the drift, there against the face was the dangling length of fuse they had lit earlier. As they approached it, the fuse suddenly jumped to life and spit forth a shower of sparks. Both men knew what was to happen within seconds and jumped between heavy timber sets as the charge detonated, spraying shot rock down the drift[25]. What had happened was the fuse hungfired after being lit and their poor timing saw them yards from the loaded heading as it resumed burning. Luckily for them they escaped severe lacerations, lost fingers and eyes, broken teeth, and broken bones by hiding behind the heavy stulls.

For the most part, prevention of misfires was under the miner's control. The causes were often a function of haste, lack of attention to detail, and a lack of training in regards to proper preparation and loading of charges. For the generation of miners that made the switch from blasting powder to dynamite in their working lives, misfires were a manifestation of learning by trial and error. Over time miners found that caps were best crimped with proper crimping tools rather than with teeth for a good seal, that fuses needed to be lashed to cartridges to prevent being pulled out, and care taken when tamping the charge in the blast-hole. Mining and explosives engineers certainly forewarned about poor priming and loading practices which would have eliminated many misfires, but again we are reminded of a communication gap between engineers and miners working underground. In addition to poor preparation techniques and faulty products, a great many misfires were caused by the manner in which explosives were treated and stored. As discussed above, poor storage practices allowed explosives to degrade to the point where they became unreliable.

Based on the above conclusions, the best approach at stopping misfires was to go to the source of the problem: miners and mining companies. Beginning shortly after 1910, mining and explosives engineers, explosives manufacturers, and the Institute of Makers of Explosives embarked on an education campaign to teach miners and their superintendents the minimal but necessary procedures for successful priming, loading, and firing of blasting powder and dynamite. As early as 1912 explosives manufacturers enclosed pamphlets on safe use and handling of high explosives in boxes of their products[26]. However many miners did not adhere to the rules and recommendations out of haste, laziness, and lack of knowledge. Still, the 1910s marked the turning point at which misfires and other explosives-related accidents declined. Ultimately the education campaign proved successful as proper methods for handling, priming, loading, and firing spread through the nation's underground workings, and accidents associated with misfires dropped off by the 1920s.

When a drilling and blasting crew reported a misfire, some unlucky miner or shift boss had the miserable task of extracting the charge before work could go on. Misfired dynamite was much more of a problem than was blasting powder, because when tamped, dynamite became a compact, putty-like mass of nitroglycerine that was highly sensitive to the slightest touch. Blasting powder, on the other hand, could be neutralized with water. There was no way to extract misfires that satisfied both mining companies and miners. Mining companies, always pressuring miners to maximize production, wanted misfires extracted as quickly as possible, but the faster the extraction, the greater the risk. It comes as no surprise that, given the air of corporate dominance over labor in the mining industry prior to the 1920s, miners had to extract misfires according to company policy: expeditiously.

Mining and explosives engineers recommended several measures of mitigating misfires that minimized the risk, but these were time-consuming and railed against the policies of mining companies. One of the recommended methods involved drilling holes around the missed charge, blasting it out, and searching the shot rock for it. Another method involved ever so cautiously picking the stemming material out of the drill-hole, inserting a dynamite cartridge on top of the misfire, and detonating the fresh charge and the misfire at once[27].

Many mining companies felt that drilling a ring of holes around a misfire, blasting it out, and searching for the old charge was much too time-consuming. Company superintendents and shift bosses felt that inserting extra charges into the drill-hole over the misfire and shooting what amounted to an abnormally large charge would ruin the working face for the next round, in addition to consuming time for the cautious removal of the stemming material. Most miners, with encouragement from their danger-numbed peers and impatient mining companies, preferred to completely gut the old drill-hole, reload it with fresh dynamite, and shoot it. What many miners failed to understand was how very sensitive a well tamped charge of nitroglycerine-steeped dynamite was. In Carbon County, Pennsylvania, on August 17, 1926, a coal miner who did not wish to spend time dealing with a misfire according to the recommendations of engineers attempted to drill out the entire mass with his coal auger. He successfully bored through the stemming material, and when the auger's teeth disturbed the charge, it detonated, blowing him up[28]. In coal mines this method of retrieving misfires was popular, especially among contract miners paid by the ton. In softer material where the drilling was easy, it was ludicrous not drill holes around missed charges and blast them out, yet due to laziness, numbness to danger, and ignorance, miners attempted to take the shortest route.

In the cases where hardrock miners attempted to totally gut a misfire, they had to use fairly aggressive action to scrape out the densely-compacted dynamite charge. In the Piute Mine near Goodsprings in southern Nevada, an old-time hardrock superintendent simply named "Sully" was using the old tried and true method of scraping and cleaning the charge out with a drilling spoon. All was going well when the day shift, on the surface eating lunch, felt the ground shake. They figured that Sully had successfully blasted out the missed hole and that he would shortly be up for lunch. Several minutes went by and there was no Sully. The superintendent then went down to find Sully, and find him he did. Sully had managed to successfully set off the misfire, but all too soon, for the drilling spoon had been shot into his chest[29]. Sully was one of many unlucky miners who paid the ultimate price of using a little too much force in picking and scraping out a misfire. Too many times miners watched each other successfully extract a misfire, unaware of just how close to the edge they were.

The only truly safe way of mitigating dynamite misfires was long in coming. It consisted of a high-powered water jet to wash the stemming and dynamite out[30]. Unfortunately, mining companies rarely employed this method because it required a tremendous amount of water, which many mines had only limited access to, and considerable capital was necessary to buy and assemble the plumbing and high-pressure pump. The lack of water and the lack of capital excluded most mines.

Under any circumstances misfires were dangerous and time-consuming to deal with, hence prevention proved the best solution. By the 1910s miners had learned through experience, word of mouth from other miners, and through information disseminated by explosives makers, engineers, and the Institute of Makers of Explosives the best way to avoid misfires was through proper storage, handling, priming, loading, and firing.

As we have seen, explosives brought to mining tremendous benefits, and they also had significant drawbacks that miners were forced to confront. The production of gases proved to be one of the most insidious problems. Gas byproducts were unavoidable, because explosives functioned when their solid ingredients converted to gases during reaction. When properly used, explosives filled mines with many inert but non-oxygenated gases such as carbon dioxide, nitrogen, and hydrogen. When coupled with miners' respiration, the heavy use of flame lights, and naturally occurring gases, explosives' byproducts displaced oxygen to the point where miners' candles barely sustained a flame, and even went out in some work areas. But the greatest problem lay in the poisonous gases produced by old, degraded, and poorly manufactured explosives.

Many miners and mining companies were not aware of the connection between the production of gas byproducts, and the poor storage and handling practices which subjected both dynamite and blasting powder to degradation. Improper storage exposed dynamite and blasting powder to moisture, and allowed dynamite to freeze or overheat, so that they produced asphyxiates such as carbon monoxide and outright poisons such as hydrogen sulfide, carbonic oxide, and nitrous oxide. Imagine the severity of the problem in large mines where many miners were blasting, in some cases improperly or inefficiently due to haste, with little or no artificial ventilation to replenish fresh air.

Dangers created by gases were both immediate, and apparent only after prolonged exposure. The immediate effects of explosives' gases could be life-taking, as illustrated by a coal miner in Northumberland County, Pennsylvania, in 1927 who shot his rounds and returned to the breast to investigate. The blast had displaced all breathable air and he suffocated and tumbled into a loading chute[31]. Gases also had a subtle effect by harming miners over the course of time, tragically in some cases.

Dominic McElhenny, a miner leaving the Marie mine near Philipsburg, Montana, in 1891, told his fellow rider in the bucket that he was getting dizzy as they passed the 200-foot level. The partner noticed that McElhenny was getting pale, and tried to tighten his grip on him as they passed the 100-foot level—when McElhenny slipped out and fell to his death 400 feet below. The coroner's jury was told that powder smoke and 'bad air' had 'pretty near knocked out' the miners before McElhenny got in the bucket to ride to the surface[32].

Blasting powder and dynamite contributed foul gases in their own ways. Under ideal conditions the combustion of fresh, pure blasting powder theoretically yielded mainly nitrogen and carbon dioxide. In addition, by nature blasting powder contained charcoal which formed carbon monoxide during combustion[33]. Blasting powder was notoriously sensitive to water and as its manufacture, handling, and storage were rarely ideal, the powder was undoubtedly subject to some moisture. Hence it comes as no surprise that when miners used powder, it fouled mine air with additional gases, including carbonic oxides, nitrous oxides, sulfous oxides, hydrogen sulfide, and carbon monoxide, all of which were either poisonous or asphyxiates[34]. Blasting with powder also produced a smoke of solid byproducts and blasted material. Mines in which the denizens of the underground consistently used blasting powder became so fouled with smoke and gases that one mining engineer complained it interfered with his vision as he inspected the workings[35]. Miners who worked day-in and day-out in this type of environment were characterized in 1862 by an editor in Central City, Colorado who claimed they could be spotted in a crowd "by the pallid countenance and haggard expression"[36].

Then came dynamite in the early 1870s. Its manufacturers claimed their new, innovative products eliminated foul gases and that it was virtually "fumeless"[37]. Nothing was farther from the truth. Detonation of pure dynamite under ideal conditions was supposed to produce carbon dioxide, nitrogen, oxygen, and water vapor. However, like blasting powder it was rarely manufactured to perfection and except for gelatin, moisture degraded it. In addition, when chemists calculated its gaseous byproducts they often failed to take into account its waxed paper wrapper and minor ingredients in its active base, such as tar, coal, carbon, cellulose, resin, and starch. Generally, heavily carbonated bases tended to give off carbonic oxides and carbon monoxide, while heavily oxygenated bases such as saltpeter, and nitrogen-based materials gave off toxic nitrous oxides, which were quite poisonous[38].

Over the course of nearly 200 years miners learned the behavior of smoke and gases generated by blasting powder; its gases were heavy, and their pungency and smoke made them somewhat tangible. When dynamite was making its way into the mining industry in the 1870s, miners expected its gases to possess similar qualities, but they learned the hard way that its gases were not as predictable. In Marquette County, Michigan a miner who was drilling and blasting a raise climbed up to the working face to inspect it after he shot his round, disregarding orders. He was probably unaware that the dynamite's gases would not settle downward as expected and when he reached the top of the ladders, he was overcome and fell forty feet to the bottom[39]. In Montana three miners ate lunch on the surface after blasting in a shaft to allow gases to disperse. When they returned to the bottom of the shaft to finish preparing a dewatering pump, they bent over to tighten some bolts and never got back up. Gaseous byproducts filled the shaft's sump, they inhaled the foul air, passed out, and drowned in the water[40]. The above anecdotes illustrate that dynamite's gases were difficult to predict. The traditional assumption that if a miner's candle burned, the air around it was breathable could no longer be relied on because even if the air held enough oxygen, it may have also held enough poisonous gases to be fatal.

Some of the factors affecting the production of gases were within the miners' control and others were beyond their reach. Certainly the quality of manufacture, time elapsed from manufacture until purchase, and treatment of explosives en route to market were beyond the miners' control. In many cases the way mining companies treated their stock was also

Figure 14. By the 1910s the explosive and mining industries acknowledged the environmental problem posed by explosive byproduct gases. The explosives industry approached the problem by promoting better products, which is reflected in this 1915 advertizement for DuPont Gelatin. (Source: Mining Press Jan. 6, 1915, p 37.)

out of the miners' control. Here storage magazines played a key role. Dynamite stored too hot for too long degraded and contributed to production of bad gases. But who would have guessed it when dynamite companies claimed it could safely be stored at temperature of up to 212° F? On the converse, dynamite froze below 50° and required thawing prior to use. In many cases miners' thawing practices were inadequate and hasty, leaving its core partly frozen, and detonating partly frozen dynamite was a poisonous gas producer second only to burning it.

Factors under most miners' control tended to relate to application. In addition to complete thawing and minimizing exposure to moisture, the method of detonation was important. Using a strong, fresh blasting cap appropriate to the product was necessary. Many miners learned through misfires and *stinkers* (charges that partially detonated and emitted foul gases) that products such as gelatin, ammonium nitrate, and nitrostarch required something stronger than the No. 3 caps used with old-fashioned straight dynamite. Most mining engineers agreed that the single best way to get dynamite to produce poisonous gases was to burn it, yet this is exactly what miners accomplished when they attempted to detonate it with small blasting powder charges instead of blasting caps. This behavior was associated more with coal mining than hardrock mining. Like storage practices, loading methods also had the potential to contribute to the gas problem. Poor tamping of charges often resulted in *blow-out shots* in which the explosives did not reach their optimal temperature and in essence burned[41].

Last, the miners had to use the best explosives for their work. For instance, if a miner had to use blasting powder in wet conditions, *A powder* was much better at minimizing gases than *B powder*. Likewise, for the choice of high explosives gelatin and ammonium nitrate were cleaner-burning than straight dynamite, and gelatin was best of all in wet conditions. Yet, most miners chose explosives based on economic criteria rather than a performance which would minimize gases. Because of their low costs, B powder remained the most popular blasting powder, and hardrock miners used straight dynamite, which was the cheapest high explosive, almost exclusively. The editor of one of DuPont's Blasters' Hand Books summed up another significant reason miners did not always chose the cleanest-burning explosive:

> It may be said that miners may be using an explosive that by no means is best suited to the purpose for no other reason that they have always used it. The prejudice in favor of a certain explosive is so great that the utmost patience and perseverance are required in order to induce miners to use an explosive which is really better adapted for the work in their particular mine[42].

Several generations of miners were forced to work in a stifling atmosphere before mining companies applied economical solutions. The problem was attacked from several angles. First, explosives manufacturers improved their products in the 1910s by revising their formulas and scientifically balancing the ingredients to minimize noxious gases. At this time miners began to favor cleaner explosives, such as gelatins, ammonium nitrates, and extra formulas over straight dynamite, which was one of the worst gas producers[43]. Explosives makers, mining and explosives engineers, and the Institute of Makers of Explosives did their share to address the problem of gases. As part of the campaign against misfires, they educated miners on storing, handling, choosing, and using explosives safely and efficiently. Some mining companies took the problem directly out of the hands of miners and hired powder monkeys to conduct blasting, hoping the explosives specialists might be more effective.

The third approach was to alter the miners' environment. Prior to the 1910s mines often relied on natural air currents to replenish fresh air, and in many mining districts companies found it in their best interest to link their individual workings to create complex circulation systems. But this was not always enough. In Colorado's Cripple Creek district it has been said a person could travel underground from the town of Cripple Creek to Victor five miles south and never see daylight. Considering the miles of interconnecting workings and number of shafts, air circulation had great potential. Yet, during the district's early years miners about to go on shift would look southwest to the Sangre de Cristo Mountains, and if clouds hung low over them, the miners knew the barometric pressure was low, gases were depressed in the mines, and they would not go down[44].

The best solution for providing miners with fresh air proved to be artificial ventilation. "It is obvious that the waiting period between the blast and the return to the face can be considerably shortened by ventilation. A good ventilating system accomplishes two purposes: first, it removes the smoke and fumes produced by the blast, and second, it assures a supply of fresh air to the men at all times"[45]. Until the 1920s ventilation technology used by North American miners ranged from primitive, passive means termed *natural*, to complex mechanical systems termed *artificial*. Many mining companies, large and small, recognized natural ventilation as effective only if a mine had two or more openings, and operations without had to institute artificial means. A primitive form of artificial ventilation used by operations with limited financing and confined to shallow workings involved erecting a canvas sail on a wood frame and ducting the air it collected into the adit or shaft through a canvas sock. In the latter portion of the nineteenth century miners adapted stove pipe and riveted steel pipe as air ducts. Another form of ventilation used in Western mining districts prior to the 1920s involved

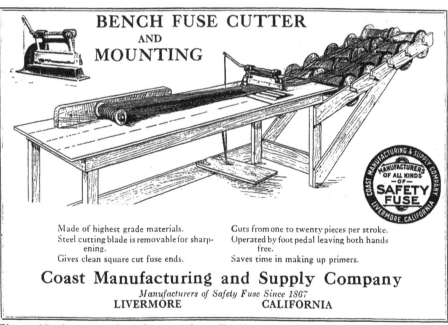

Figure 15. At most mines, large and small, miners primed their own rounds near the points of work. At a few high-production operations a powder monkey held the duty of priming the dynamite. His shop might have featured appliances such as a bench crimper and fuse-cutting table with factory-made cutter. (Courtesy Hercules Inc.; Source: Explosives Engineer Aug. 1926, p 309, Oct. 1925, p 326.)

connecting large forge-bellows to stovepipes, canvas tubing, or riveted steel pipes suspended from the mine's ceiling. The occasional pumping of the bellows forced air through the ducting into the mine workings[46].

The popularization of compressed air rockdrills in the 1890s improved ventilation in mine workings where there was previously none, especially where large numbers of the machines were employed. Although somewhat tainted with oil fumes, the drills' exhaust and leaks in the air plumbing and connecting hoses introduced fresh air. Ironically, until the 1910s the drill operators, who were miners closest to the sources of the fresh air, were last to benefit because they worked in a cloud of rock dust, another product of compressed-air drills, which gave them silicosis.

Many hardrock mining companies considered exhaust and plumbing leaks associated with rockdrills to provide adequate ventilation. Once wet drills became popular in the 1910s and the deadly rock dust was allayed, this source of air was a significant improvement for all, but in large mines it was still not enough. In the 1910s, and especially during the 1920s many large operations began employing ventilation blowers and networks of ducts to force large volumes of fresh air in and foul air out of the mine. Proper ventilation became the norm in many mines as the years went on. By the 1930s and 1940s forced-air ventilation coupled with improved behavior on the part of miners greatly improved the air quality in many mines.

CHAPTER FOUR

THE TECHNOLOGY OF OPEN PIT MINING & BLASTING

Huge deposits of industrial metal ores, such as iron and copper in the Midwest and Northern states tantalized capitalists and mining companies as early as the 1880s. The problem with exploiting the ores lay in their dilute state; they were just not rich enough to turn a profit given then-current mining and milling technology. One of the most fundamental ways to make the massive deposits pay was to lower production costs and mine then in economies of scale. While drilling, blasting, and mucking rock at ground surface did not

Figure 1. Sooty skies, the smell of coal smoke, broken earth, and the rumble of heavy machinery formed the environment of the open pit mine worker. Like in most open pit mines, in this early 1920s scene, makeshift railroad spurs were hastily laid for steam shovels and churn drills bored blast-holes on ground stripped bare of top soil. (Courtesy Hercules Inc.; Source: Explosives Engineer *Sept. 1924, p 316).*

begin to approach the cost of underground operations, mining technology still restricted the tonnage that could have been produced. But by the year 1900 four factors came together which lowered the costs of surface mining, and increased tonnages.

First, the efficiency of rock drills had improved to the point where they were capable of drilling fairly deep vertical holes in short order, provided their operators were experienced. Second, during the 1880s and 1890s, engineers adapted churn drills for boring oil wells through hard cap stone. Until this time, drilling outfits had traditionally used the machines to bore water wells up to several hundred feet deep in sedimentary ground. The result was a machine capable of drilling large-diameter nineteenth century, until by the 1890s dynamite could have been had for as little as 12 cents per pound, and blasting powder for 7 cents per pound. Mechanization and technology brought the cost of mining down enough to place low-grade ores within economic reach.

The final ingredient in the birth of open pit mining consisted of a handful of mining engineers willing to experiment by applying the mechanization in an efficient manner, and by mining companies willing to risk the capital. Engineers attempted to use mechanical earth-moving on a large scale for stripping over-burden off ore bodies, they applied rock drills or churn drills for boring deep vertical blast holes in experimental patterns, and loaded the holes with unprecedented amounts of explosives. By attempting to use unconventional and capital-intensive methods at places such as Ely, Nevada, Bingham Canyon, Utah, and Hibbing, Minnesota, a

Figure 2. The steam shovel proved to be a cornerstone of open pit mining. Developed around the turn of the century, these mighty machines moved unprecedented tonnages of earth and rock in little time, increasing ore production and reducing the costs of surface mining. The first steam shovels were mounted on railroad car chassis, and by 1920 manufacturers introduced small models propelled by caterpillar tracks. (Source: International Correspondence Schools, 1907, A. 41, p 9).

handful of mining companies set a technological precedent.

During the 1900s and 1910s open pit mining underwent dramatic growth and development in terms of technology, methods, and geography, and much of the change centered on blasting, which was open pit mining's hingepin. Mining engineers honed existing blasting methods, explosives manufacturers introduced numerous low-cost products designed to meet the needs of open pit mining, and machinery manufacturers introduced implements geared to withstand the rigors of surface work. Some of the methods developed by engineers included working ore bodies in benches by systematic drilling, blasting, and shot rock removal. Bench sizes, which depended on the rock type and the drilling equipment, ranged from 15 to 100 feet high and wide. Because nearly every ore body was unique, exact drilling patterns, hole sizes, loading methods, and choices of explosives had to be customized. As technology and methods improved through the 1910s and 1920s, open pit mining expanded throughout the nation as companies extracted metals such as gold, silver, copper, iron, molybdenum, and mercury, and minerals including limestone, gypsum, salt, coal, and road ballast.

Despite ongoing technological advances, the basic tenets of open pit mining remained unchanged. To turn a profit, mining companies had to handle enormous quantities of rock with minimal expenditure of time and money. To meet this goal, mining engineers adopted the practice of drilling deep blast-holes capable of containing huge explosive charges, arranging them in patterns that covered, at times, acres of ground, blasting thousands of tons of rock at once, and mucking it out with huge shovels. As the costs of underground mining rose during the twentieth century and high grade ores became depleted, the popularity of open pit mining increased, until the number of open pit operations eclipsed underground mines.

Drilling Blast—Holes

While open pit operations can be described as mining on a grand scale, it shared with the underground segment a cycle based around blasting. Pit crews drilled blast-holes, loaded them with explosives, shot the round, and removed shot rock. Because huge volumes of rock had to be fractured in single blasts, drilling technology and drill-hole patterns played a significant role in the economic feasibility of open pit mining. Efficiency in drilling came about from mining engineers who tried to adapt underground machinery and methods, and the drillers who field-tested the applications. Some technologies and methods crossed over well from underground mining, while others flopped.

As a result of calculation and experimentation, the open pit industry employed perhaps a greater variety of drilling methods and machines than any other type of mining. Hand-drilling quickly proved inefficient, but piston and hammer rockdrills saw extensive use. Mining engineers were quick to embrace new drilling technologies, which for the first several decades of the twentieth century included drill trucks, wagon drills, churn drills, and air-rotary rigs beginning in the 1950s, in hopes of increasing production and reducing costs.Drilling technologies may be divided into three categories: rock drills, churn drills, and air-rotary rigs. Each group was technologically unique, and open pit mining methods had to be adapted to their specific capabilities.

The wide-spread acceptance of reliable, reasonably fast piston drills in underground minds and stone quarries in the 1890s formed one of the cornerstones of open pit mining. These early drills proved themselves capable of drilling deep holes in much less time than hand drilling, which resulted in lower costs of mining and increased production. However, the piston drills used in underground mining had to be adapted to the requirements of open pit mines. First, the sources that powered the drills had to change. Drills used underground ran on compressed air or steam, both of which were plumbed into the mine. Steam was generated in boilers, and stationary steam-driven or electric compressors provided air. Both of these power sources had to be mobilized for open pit mining, because the extensive size of pits and the

Figure 3. Mining engineers and experienced quarry workers brought the quarry bar for mounting rockdrills from stone quarries into open pit mines by the 1910s. Quarry bars, developed in the mid-1880s, ranged from simple sawhorses constructed of heavy pipes to special devices with mechanized tracks for drill mounts, as illustrated. The early lithograph also depicts a piston drill, which were not replaced for surface work until the 1920s. (Source: Kirk, Aurthur, 1891, p 89).

huge blasts made laying air plumbing from a remote station inefficient. Gasoline and steam-powered upright and small straight line compressors proved most adaptable, and they were mounted on wheeled frames. In addition to modifying power sources, drill mountings had to change.

In the 1890s underground mines and dimension stone quarries favored two methods of mounting rockdrills, which were the column and the tripod. The column was restricted to use underground, because it required a ceiling so it could have been screwed into place. The tripod, however, found ready application for surface work, but it was not without limitations. Tripods were made of heavy steel and had fully adjustable legs hinged to a saddle for mounting a drill, but they could mount only one machine. To increase the efficiency of drilling, the stone quarrying

Figure 4. The transition from piston rockdrills to hammer drills for boring blast-holes in open pit mines lasted from the 1900s into the 1920s. Many mining companies introduced hammer drills into their operations during the 1910s as drill makers improved the technology. The illustration depicts a drilling crew operating a hammer drill mounted on a tripod during the 1910s. The steel-spiral-wrapped air hose, air pipes, and the two buckets containing tools and drilling water are accurate. (Source: E.I. DuPont de Nemours & Co. 1920, p 55).

industry (discussed in detail in Chapter 5) spurred the development of several alternative mounts. One of the best, appropriately named the quarry bar, was a heavy steel tube up to sixteen feet long on four or six legs, and it served as a stout steel sawhorse capable of mounting over four drills, while remaining steady under severe vibration. Both the tripod and quarry bar saw extensive use in open pit mines. The tripod's portability and the independence it afforded made it efficient when large hole patterns were being drilled on uneven terrain. When a tight, rigidly defined hole-pattern was necessary and the engineer wished to have multiple drillers working in a small area to expedite blasting, a quarry bar answered.

In the early 1900s the hammer drill arrived underground, and by the 1910s it had proven itself better than the old piston drills and began to replace them. However, in open pit mining piston drills remained the favorite until the early 1920s because they were unrivaled for drilling deep vertical holes[1]. Drilling rates for piston drills varied according to the rock type being drilled. For instance, a piston drill could sink a vertical two-foot hole in limestone in six minutes, eight minutes in sedimentary rock, ten minutes in quartzite and pegmatite, and 12 to 16 minutes in hard granite and metamorphic rock[2]. However, real drilling time was longer because variables such as spooning drill-cuttings out of the hole, oiling the drills, and changing drill-steels every two feet also had to be included. On average, a crew of two working in sandstone with a piston drill could finish 96 feet of hole per shift, 48 feet in granite, and in dense metamorphic rock 36 feet[3].

By the early 1920s the open pit mining industry whole-heatedly accepted hammer drills. Hammer drills manufactured in the early 1920s were faster than the older piston drills, and they had several time-saving features, such as lighter drill-steels that simply slid into their chucks, better cutting-bits, and water or air forced through a hollow steel that flushed cuttings out of the hole. By the 1920s a hammer drill of the drifter type could drill 200 to 250 feet of hole per shift in soft rock, which should be compared with piston drill rates[4].

As open pit mining processes improved between the late 1890s and early 1920s, rockdrill mountings evolved from the quarry bar and tripod into expensive, complex vehicles. While they cost mining companies more capital, the vehicles sped the drilling process and eased the backbreaking work of manually moving 350 pound rockdrills, tripods, and 35 foot-long drill-steels. The first break away from a legged frame was the gadder. Early gadders, most having been custom made, consisted of modified quarry bars mounted on rail axles. Rail gauge varied from the standard 18 inches used in most underground mines to six feet, but within a decade most open pit mines and stone quarries used standardized, factory-made axles. By the 1910s factory-made gadders typically employed large piston or hammer drills mounted on a very short boom or T-bars on a baby-gauge rail car. One of the gadder's main benefits was that the drill had a great length of travel, allowing use of longer drill-steels, lessening the number of times they needed to be changed. In addition, in only minutes a drilling crew could have broken down, moved, and set up the unit for drilling, while it took longer to do the same with a tripod. The drill truck, invented in 1915, combined the convenience of the gadder with greater mobility. A boom with a drill was mounted to the rear of a heavy-duty chain-driven truck chassis, instead of a rail car. Once the

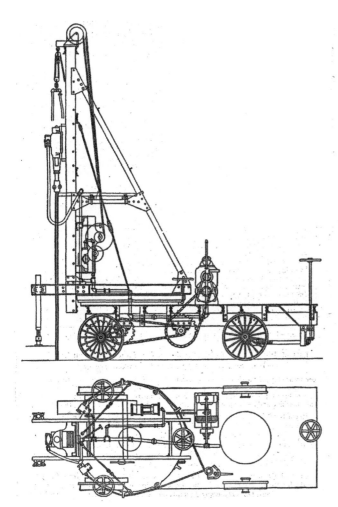

operator raised the boom, one of the crew members lowered the drill toward the rock by means of a hand-cranked winch. More maneuverable yet, some driller or engineer had the bright idea mounting a 16-foot boom and drill on a small two-wheeled frame trailer. Invented in 1924, this contraption, known as the derrick drill, served into the 1950s. The derrick drill was almost as portable as the old-fashioned tripod, it was as easy to set up as a gadder, and it cost less.

Figure 5. The illustrated engineers' elevation and plan drawings depict typical circa 1910s and 1920s drill trucks. The truck features a heavy drifter hammer drill mounted on a boom, which rests on a turntable. A small gasoline engine drives an air compressor for the drill, and it powers the truck via chain linkage. Because of mobility and little set-up time, drill trucks were well suited for boring numerous blast-holes in rapid succession. (Source: Peele, 1918, p 185).

The open pit mining industry constantly demanded better, more efficient machinery. Manufacturers responded, and in the mid-1940s a drill maker introduced the Jumbo, which was the first self-contained mobile drill rig capable of supplying its own air. The jumbo used above ground should not be confused with the rig by the same name used underground. The jumbo used for open pit mining had its own drill-steel rack, an air compressor, and a monster hammer drill, all mounted on a caterpillar chassis. The jumbo proved to be quite an effective rig in open pit mines because of its mobility, its durability, and its heavy drill could quickly bore six to eight-inch diameter

Figure 6. The open pit mining industry embraced wagon drills during the 1920s because of their great mobility, fine performance, and low cost. By the 1940s wagon drills became commonplace for a variety of surface blasting projects. The scene illustrated depicts laborers using wagon drills to bore lifter holes during the early 1940s. A fiberboard box of detachable drillbits has been placed on a boulder in the lower left. (Courtesy Hercules Inc.; Source: Explosives Engineer *Jan.-Feb. 1944, p 2).*

holes. The jumbo proved to be such an efficient design that it is still used in open pit mines today.

In addition to employing ever-faster, larger drills, open pit mines also pursued a trend of lighter, smaller models. From its inception, open pit mines cherished the hand-held sinker drill, developed by Ingersoll-Rand in 1912. By the 1920s, leading drill manufacturers such as Ingersoll-Rand, Sullivan, Cleveland, and Cochise produced heavy models, around 55 pounds which could drill vertical holes 15 to 20 feet deep, enough to satisfy some open pit needs. With a heavy sinker one pit worker could drill up to 150 feet of drill-hole during his shift[5]. Open pit mines commonly employed lighter sinkers that workings used to drill starter-holes, pop-shots, and small blast-holes.

The choice of drilling technology relative to its era of popularity is indicative of an open pit mine's financial and engineering state. Well-capitalized mines with progressive engineers willing to try new technologies tended to use expensive, innovative machines before they became common. In addition, progressive open pit mining companies equipped themselves with a variety of drills, from churn units to handheld sinkers because each type was most cost effective when performing a particular task. On the contrary, mining companies with conservative management and limited finances relied on proven technology and a narrow array of machinery.

Like underground mining, blasting in open pit mines could only be effective if drill-holes were laid-out in a pattern. By nature, open pit mining companies worked mineral deposits with methods similar to the underhand stoping practiced underground. In open pit mines, engineers applied three basic hole patterns in conjunction with rockdrills to perforate a bench with holes. The two most effective arrangements included a grid, a checkerboard pattern, while a third, the semi-circular pattern, worked best in hard rock. Density of the rock governed the spacing between drill-holes. In soft rock, drillers spaced their holes 10 to 20 feet apart and 10 to 20 feet back from the edge of the bench. In very hard rock, such as granite, drillers spaced holes as tightly as five feet apart. A tight pattern fractured the rock finely, which some engineers and mine bosses preferred because fine shot rock was easier to handle and less damaging to equipment[6]. Some pit conditions, such as high benches and dense rock, required holes drilled horizontally into a bench's toe, which assisted the pattern, in essence acting as lifters.

Workers typically drilled holes between 15 and 30 feet deep, and this involved aggressive labor. For example, changing drill-steels was an onerous task. Like underground mining, starter steels for drills were approximately a foot long and successive lengths in the set of steels increased one to two feet per change. Trying to change 20 and 25 foot drill-steels, and cleaning all of the cuttings out of a deep drill-hole with a 30 foot-long drilling spoon was no easy task.

Engineers and pit bosses not only adapted drill-hole patterns from underground stoping to open pit mining, they also applied variations of firing orders for blasting. Often, mine workers arranged the charges to fire a cut round first, followed by the rest of the holes, either row by row, individual holes, or small groups of holes in sequence.

In the interest of producing ore in economies of scale, open pit mining companies applied churn drills for drilling large diameter blast-holes. Cumbersome and seemingly inefficient, churn drills revolutionized open pit mining because with them, pit crews bored holes capable of containing previously unheard-of quantities of explosives, which resulted in more rock brought down in bigger blasts than was possible with with rockdrills. Churn drills, originally known as well drills, had been employed for drilling water and oil wells at least as far back as the 1870s. During the 1880s and 1890s, engineers and drill operators retooled the machines and gave them more power to facilitate the search for oil in deep sediments and under sedimentary cap rock. By approximately the year 1900 the technology of churn drills had advanced to a state where the awkward machines were effective at drilling large holes in both fairly hard and soft rock.

Around the time churn drill technology permitted work in relatively hard rock, a few visionary mining engineers and mining companies willing to risk the capital attempted to apply the machines to open pit mining. Most mining companies experienced success, and other outfits followed suit, and in response churn drill makers began developing specialized hardware and drill-bits for the industry.

Churn drills saw extensive development and adaptation between their introduction in the 1890s and their widespread embrace in the late 1910s. Up to the early 1920s, most churn drills consisted of a stout chassis with a tall wood mast at one end, winches in the center, and a power plant at the other end, all on heavy wheels. The winches were intended to manipulate utility lines hanging from the mast, and to raise the mast, which stood up to 50 feet high. Older and less-expensive churn drills were horse-drawn while more sophisticated rigs were self propelled. Into the 1920s power plants included either donkey engines supplied steam from upright boilers, or petroleum engines. By the 1920s rig makers offered electric-powered models, and gasoline and diesel remained popular, while steam became obsolete. Churn drill rigs were by no means self-contained; crews had to lug several support trailers to carry tools, bailers, cable, and other equipment. Drills came in various sizes, from light-duty rigs weighing six tons or less, to heavy-duty rigs weighing up to 22 tons[7]. Originally, the chassis and masts were made of heavy timbers, but in the 1900s manufacturers began offering heavy-duty rigs based on steel framing.

A string of tools constituted the business-end of a churn drill. The assemblage consisted of a hardened steel cutting bit screwed to a drill-stem, which was threaded onto a fitting shackled to the hoist cable. The cable passed over a pulley atop the rig's mast and wound around one of several winches mounted in the rig's center. Drilling action, called spudding or churning, was similar in action and effect to an old-fashioned mortar and pestle. The machine jerked the string of tools up about four feet and let it free-fall, and the crashing bit mauled, chipped, and broke the rock. A walking beam, driven by a reciprocating cam that rotated near the rig's center, pushed on the cable behind the mast, causing the string of tools on the far side to rise. Drillers found that a rate of 50 to 60 times per minute to be the fastest cycle without snapping wood beams or damaging the cable[8].

The rock bits used for churn drilling resembled those used in conjunction with compressed air-powered rockdrills found underground; churn drill bits featured either a chisel blade for cutting, or a star bit for cutting and mauling

Figure 7. While churn drills came from a variety of makers in several sizes, the basic form remained constant. The machine consisted of an engine that drove cable drums and a reciprocating walking beam, a mast at one end, and a string of "tools" attached to a cable. The walking beam jerked the string of tools up and let it freefall repeatedly, drilling a hole in rock. The illustrated churn drill has been set up for use, the string of tools leaning against the mast, which is guyed with ropes for stability. The drill captured in the photo is a relatively small tractor-drawn unit powered by a gasoline engine. (Source: Gus Pech, 1927, p 12).

the hole bottom. Most chisel bits had concave flutes which flushed drill-cuttings up and away from the rock surface. Engineers based their choice of bit size, weight, and shape on the type of rock being drilled[9]. Similar to the way in which sets of drill-steels for rockdrills decreased in diameter with hole depth to prevent them from jamming, the bits used for churn drilling also decreased in diameter. Then a bit jammed in the hole, the drilling crew held little hope of retrieving it, despite expensive, custom-made fishing tools optimistically kept on hand by open pit mining companies. The engineer based his choice of the bit's size and weight on the size of the rig. Small rigs required lighter strings of tools than large rigs. In general, pit crews used starter bits that were between 10 to 16 inches in diameter, and they used finishing bits between 4.5 and 6 inches in diameter. The bits alone often weighed hundreds of pounds, while drill-stems could have weighed in the thousands of pounds. As a unit, the entire string of tools for a light-duty rig weighed less than 1,500 pounds, while the string used on heavy rig weighed up to 5,000 pounds[10].

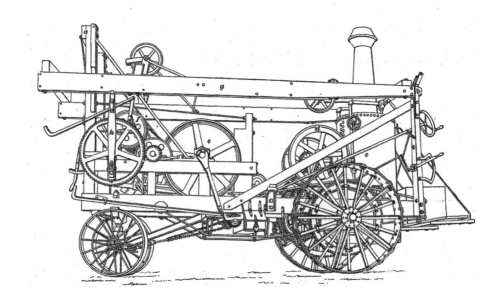

Figure 8. Many churn drills early in the century featured steam engines powered by small upright boilers. Large units, such as the model illustrated were also self-propelled. To prepare a churn drill for travel, the crew disassembled the mast, fastened the cables, and stored the tools. Churn drills crept along at an extremely slow pace, often towing trailers loaded with equipment. (Source: International Correspondence Schools, 1907, A. 34, p 29).

Pit crews found that churn drill holes quickly became fouled with rock cuttings, which interfered with the bit's progress. Since it was not practical to spoon cuttings out of a hole 60 foot deep, as did miners drilling shallow holes with conventional rockdrills, drilling crews devised an alternate means of extracting the rock dust. One of the driller's helpers dumped water into the hole while the drill operated, and the tools' churning action kept the cuttings suspended in the water. At a certain point, the driller raised the string of tools out of the hole and quickly sent down a special water bailer on a separate cable, known as a sand pump. The bailer collected the mud through a one-way valve, and the driller brought it up and dumped it out. The crew repeated the process until they had bailed the hole as dry and free of cuttings as practicable.

Open pit mining companies wanted crews to drill the maximum number of holes in minimal time. However, the rate of drilling blast-holes was a function of the type and size of the bit used, the weight of the tools, and the type of rock being drilled. On average, within an hour a rig with a 6 inch bit and 1,500 pound string of tools could drill a hole 1 foot deep in granite, 5 deep in sandstone, 5.7 feet deep in porphyry, and 8 feet deep in limestone[11].

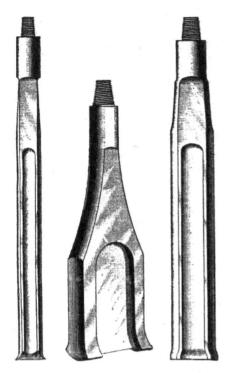

Engineers found that holes between 20 and 40 feet deep were most effective for blasting moderately dense to hard rock. The capabilities of such an unprecedented explosive charge to blast rock was staggering, and mining engineers realized the concepts of the hole patterns associated with rockdrills, which was the cenventional drilling technology of the day, were no longer applicable. In the early years of open pit mining, engineers collectively applied calculation and experience and came up with several broadly applicable patterns, which are still used today. Pit bosses and engineers found in most cases that patterns comprised of enlarged checkerboards, grids, and single rows of holes were best. Naturally, engineers and blasters found the firing orders used with rockdrill holes adapted well, which usually went row by row.

Figure 9. The bits used by churn drilling crews in open pit mines varied in exact design, but all were similar to the three illustrated. The bits at left and right were for drilling blast-holes in rock and the center bit was for boring through soil and loose ground. All bits featured tapered threads on top that screwed into heavy steel drill-stems, forming a portion of a string of tools. (Source: International Correspondence Schools, 1907, A. 34, p 16).

During the 1950s the rotary drill rig succeeded the churn drill for boring large-diameter blast-holes in open pit mines. The type of rotary rig used in mines evolved from

Figure 10. The lithograph shows a drill-stem, which provided the string of tools weight, and acted as a coupling between the bit and the rope socket. Stems went by various names and took several forms, depending on the type of rope socket and bit. (Source: International Correspondence Schools, 1907, A. 34, p 15).

large oil-well drills, developed in the 1930s. During the next ten years engineers attempted to use down-scaled, mobilized versions for drilling blast-holes in open pit mines, with limited success. Rotary rigs drilled holes of similar dimensions as those made by churn drills, but in a fraction of the time, with a fraction of the crew, with less maintenance, and they were easier to set up and break down. Because of superior performance, these durable but complex machines experienced immediate popularity, and today they are a mainstay in open pit mining. While the superior performance of rotary rigs was better for the open pit mining industry, drilling crews eyed the new labor-saving machines with suspicion.

The rotary rig drilled with a *Hugh's Bit,* which was an assembly of three rollers impregented with hardened steel teeth. The bit was coupled to a series of drill-stems threaded onto a power take-off mount at the top of the rig's short boom, and it rotated at high speeds. To flush drill-cuttings out of the hole, compressed air or water was forced through the stems and into the bit under pressure. Most rotary rigs were fully self-contained with an air compressor, a power plant, and storage for bits and stems.

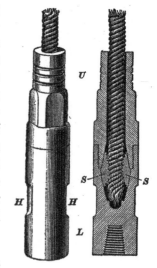

Figure 11. The rope socket, illustrated by the lithograph, formed the first link for the string of tools, which also included the drill-stem, bit, and possibly several other fittings. (Source: International Correspondence Schools, 1907, A. 34, p 14).

Until 1952 the rotary rig could only drill soft sedimentary rock because of the bit's limitations. Nothing had been developed that could withstand the spinning, grinding action in hard rock. Even in soft rock the bit held drilling progress back to 100 feet per shift. But in 1952 engineers developed a new Hugh's Bit studded with tungsten-carbide teeth capable of cutting hard igneous and metamorphic rock, and it increased the drilling rate in soft rock to 400 feet per shift. Including a set-up time of only several minutes, rotary rigs were capable of drilling a 20 foot blast-hole in 30 minutes[12]. Once the speed and dependability of these machines had been demonstrated, open pit mines ordered them by the flatcar load, quickly phasing out the comparatively dinosaur-like churn drills.

Shooting a Round In the Pits

According to outward appearances, open pit mining shared with underground mining basic elements of basting. True, open pit mining engineers and pit bosses had to select the most appropriate type of explosive,

and pit crews primed it, loaded it, and shot the round. But in reality underground practices had been adapted to the different world of open pit mining, and in many cases to the unique conditions of individual mines. Similarities existed between loading and shooting holes made by rockdrills in open pit mining and underground mining, but pit bosses and engineers threw much blasting convention out the window when it came to blasting with churn drill holes.

Open pit mining companies, like their underground mining counterparts, had to choose from a variety of blasting products to supply their crews. Pit bosses and engineers attempted to balance cost against performance. They opted for the least expensive explosive, but one which would break ground in their pits most effectively. This goal was at times difficult to achieve, because blasting conditions varied widely between open pit mines, and the comparative standards enjoyed by under-ground miners were not as cut-and-dry.

Figure 12. Churn drilling created a huge quantity of rock cuttings that had to be removed from blast-holes at regular intervals. Because churn drill holes were deep and large in diameter, the best device for removing cuttings was the sand pump, *one type of which is illustrated. The drilling crew dumped several buckets of water into the hole during drilling, they pulled up the string of tools, and lowered the pump down the hole until it rested at bottom. In a quick motion they hoisted the pump, and the upward action pulled the plunger* b *through the pump shell* a, *drawing in wet cuttings. A ball valve located in the pump's nose prevented the cutting slurry from draining out. Once at the surface, the crew dumped out the pump and repeated the process. (Source: International Correspondence Schools, 1907, A. 34, p 25).*

Still, enough similarities existed among open pit mines in regards to particular rock types, other geologic conditions, and mining methods that pit bosses and engineers were able to determine basic constants regarding choice of explosives, and how they should be loaded.

Economic factors aside, pit bosses and mining engineers selected their explosives according to performance in the type of rock being blasted. For most situations, low-velocity explosives, such as ammonium nitrate, low-percentage strength straight dynamites and extra formulas, railroad powder, and blasting powder were ideal because they cost least and performed best. In underground mining, stronger, faster explosives such as high percentage straight dynamite, gelatin, and extra formulas had to be used to blast the same types of hard rock shattered by the slower explosives used in open pit mines, because the shear size

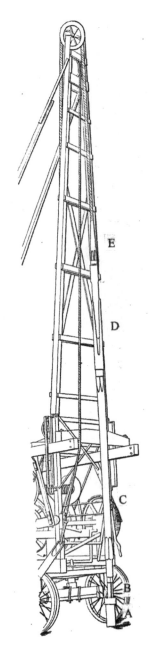

Figure 13. The line drawing depicts a complete string of tools suspended over the collar of a large-diameter blasthole. A is the hole collar, B is the bit, C is the drill-stem, D is a jar which moderated the force of the crashing blow and jerk upward, and E is the rope socket. Behind the string of tools stands the churn drill mast. (Source: Engineering & Mining Journal *Dec. 21, 1907, p 1152).*

Figure 14. By the 1930s the types of churn drills manufactured for open pit mining had come a long way. Some were powered by compressed air or electricity, but most featured gasoline engines. Larger models consisted of steel booms fitted onto steel chassis, the drums and engine were enclosed in shrouding, and the machines were propelled by caterpillar tracks. The illustration depicts a drillers' helper standing next to a string of tools, and he is probably preparing to bail the hole of rock cuttings. A sand pump hangs from a cable right of the laborer. Note that the cable for the string of tools passes over the large sheave atop the mast, while the sand pump's cable passes around the smaller pulley. (Source: E.I. DuPont de Nemours & Co. 1934, p 92).

of surface blasts exerted sufficient force. Engineers and pit bosses used costly high-velocity explosives for hard igneous and metamorphic rock because their explosions were quick and shattering. Most of these types of explosives fell into the categories of straight dynamite, gelatin, extra formulas, TNT, and nitrostarch. Because all blasting in an open pit mine was on the ground surface, the choice of explosives was not restricted by the quality of fumes and gas byproducts, as it was in underground mining where air quality was of prime concern. While the use of straight dynamite went into decline in underground mining in the 1910s because of its propensity to foul mine air, it enjoyed popularity in open pit mines into the 1950s because it was inexpensive.

Pit crews preparing and loading charges in open pit mines handled up to several tons of explosives for a single blast. Rather than fumble with thousands of dynamite cartridges, as did their underground mining brethren, pit bosses and crews alike found that explosives packed in bulk, free-flowing form saved much time in handling. The fact that explosives makers had always sold blasting powder by the keg, and railroad powder in 6 1/4, 12 1/2, or 25 pound waxed paper bags packed in 50 pound wood boxes, only fostered the practice of pouring explosives into drill-holes like liquid. But not until the late 1920s did explosives makers develop bulk, free-flowing dynamite, which was sorely needed in the pits. Once on the market, open pit mines consumed it by the box car load.

Despite the utility of bulk explosives, many situations caused by a high water table and geologic structure necessitated the used of old-fashioned, tough, wax-paper cartridges. Explosives manufacturers offered almost any diameter either as standard packaging or by special order. Typical lengths of cartridges used in rockdrill holes ranged from 8, 10, 12, to 16 inches, while 16 and 24 inches were typical of those used in churn drill holes. The most common diameters for cartridges used in holes bored by churn drills, air rotary rigs, and jumbos were 3, 4, 5, and 6 inches. In some cases cartridges expedited loading large-diameter and deep drill-holes, and they certainly protected the blasting agents against moisture. The choice of a cartridge's size hinged on matching the diameter of the drill-hole.

Loading Drill-Holes in an Open Pit Mine

The underground and open pit mining industries were completely dependent on blasting, and loading explosives into drill-holes was a necessary part of both forms of mining. In cases where open pit mines employed rockdrills, pit crews adapted the loading methods used by hardrock miners underground. However, pit bosses and crews had to experiment with new practices when loading high-volume churn drill holes. Even the conventional practices for loading rockdrill holes did not go unmodified. Hole diameter and depth, the type of rock, and the ignition or detonation device were all factors that determined exactly how pit crews loaded drill-holes.

Open pit mine crews loading rockdrill holes in hard rock adapted the procedures used by underground miners. A pit worker made primers, another worker slit cartridges to allow them to expand and he dropped them into the drill-holes, while a third crew member used a tamping rod to pack the mass down to eliminate air spaces. For blasts containing large numbers of holes, pit bosses and engineers often primed the charges with electric caps, including delay-action units to ensure a firing order. By the 1920s explosives makers offered up to 10 different delay caps and blasting machines capable of firing 100 shots for large blasts. Where the number of holes was few, or the threat of severe electrical storms severe, pit bosses and powder monkeys had the charges primed with standard blasting caps and safety fuse.

In soft rock engineers found that it was best to chamber or spring rockdrill holes prior to final loading. Chambering a hole was a way of loading and shooting a series of small charges at the hole bottom to create a cavity capable of holding more explosives than would have otherwise been the case[13].

The first stage of springing usually involved a worker who dropped one to three cartridges of dynamite to the drill-hole bottom and detonated them without tamping the hole shut. After the charge detonated, he enlarged the

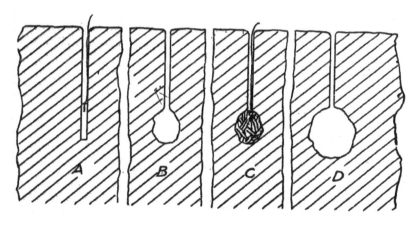

Figure 15. The basic sequence open pit crews followed for springing rockdrill holes is outlined by the illustration. First, they detonated several cartridges of gelatin at the bottom of a blast-hole. Next they inserted numerous cartridges of straight dynamite or gelatin in the enlarged chamber and detonated the charge, which created a void capable of containing enough explosive for an effective blast. (Source: E.I. DuPont de Nemours & Co. 1932, p 89).

small cavity with eight to ten more cartridges, and again with up to 60 cartridges (14). In between blasts, a pit worker probed the drill-hole with a long wooden pole to clear it of debris. The exact number of times the hole was sprung depended on the density and resistance of the ground. In very soft ground, the hole was sprung as little as one time, while in dense ground, the hole was sprung three or more times.

Engineers realized that high velocity dynamites, such as gelatin, were considered to be most effective for springing because they pulverized the rock at the hole bottom and pushed it back. Generally, open pit mining engineers recommended that the drill-hole be allowed to cool for at least a day in between charges to minimize the danger of accidental ignition of the new charges[15]. After the hole had been chambered, the pit crew usually loaded it for a final blast with blasting powder, although they occasionally used bulk or cartridged dynamite. As the crew loaded the explosive, they used a long pole to swish it into all of the nooks and crannies of the chamber. Once they had ensured the chamber was full, the crew dropped in a primed cartridge, tamped the hole closed, and prepared for the final blast.

The development of practices for loading the deep, large-diameter holes bored by churn drills was not completely unguided and experimental. Dimension stone quarry workers had worked with bulk blasting powder poured into vertical holes for decades, and oil workers had established some basic methods for working with churn drill holes. But the creation of functional practices for loading and blasting in pits was up to pit bosses and engineers.

The availability of bulk explosives made the process of loading churn drill holes easy. When the blasting crew loaded a bulk explosive such as blasting powder, railroad powder, or free-flowing dynamite into a dry churn drill hole, they simply poured it straight in. However, if there was standing water, then they had to load large-diameter dynamite cartridges up to the water line to form a dry foundation for the bulk explosive. After half of the powder had been poured, they primed the charge with a dynamite cartridge, which acted like a giant blasting cap, and they poured the rest of the bulk explosive on top of it.

Like underground mining, open pit mines presented crews with a variety of safety hazards. Danger lay in working with high volumes of explosives, and pit crews were wise to exercise great care when handling bulk blasting powder. The Nevada Consolidated Copper Corporation operated one of the nation's largest open pit copper mines near Ely, Nevada, and like other operations, the company shot large charges of blasting powder. Some of the blasting crews became a little sloppy pouring the powder into drill-holes, and on several occasions they allowed considerable amounts of loose powder to accumulate around the hole collars. Open pit mining engineers strongly advised against such clumsiness because exposed accumulations of highly flammable explosives were disasters waiting to happen. Despite the warnings, and going against common sense, the poor loading practices of Nevada Consolidated's crews continued. As a result, two gigantic explosions blew up blasting crews, and in one disaster eight workers were consumed in one ball of fire[16]. Because loading huge quantities of blasting powder was an extremely sensitive operation, open pit mining engineers recommended that all churn drills, locomotives, steam shovels, and other internal combustion driven or sparking machinery be removed or shut down.

In the late 1920s the Hercules Powder Company developed a modified blasting powder sold under the brand name of Herco. Whereas standard blasting powders were of well-sorted grain-sizes, Herco was a blending of different ones. While using powder of mixed grains was taboo in underground mining, it was found to be very effective in blasting high-volume drill-holes. Mixing grain-sizes maximized the powder's density, and when the small grains exploded first, they forced the larger, still-combusting grains into the fissures that developed in the rock during the explosion[17]. Their combustion added duration to the explosion and spread its effects. The key to efficient blasting with Herco was priming the length of the hole with primacord, which initiated simultaneous combustion of the entire powder column. Primacord, also known as quarry cord, cordeau, and shot cord, was similar in appearance and construction to safety fuse, but instead of a gunpowder core, it had a powerful, high-velocity primary explosive that detonated at the rate of 17,500 feet per second[18]. The advantages of using primacord were that it served to detonate dynamite in place of blasting caps, and when linked to all of the drill-holes in a pattern, exact firing orders could have been achieved without using electric delay systems.

When pit crews used primacord to prime the charge in a churn drill hole, they used it to aid loading the high explosive charges. The blasting crew pierced the first cartridge to be sent down the hole from end to end and passed the primacord through it. A worker tied a knot in the end of the primacord and lowered the high explosive cartridge away. Once the first cartridge was at the hole bottom, the crew pierced the other cartridges, threaded them onto the prima-cord, and sent them sliding down[19]. By using the prima-cord as a guide, the cartridges were not as likely to broach in the hole.

Blasting crews primed cartridges with blasting caps and safety fuse, and with electric caps in a manner like underground mining. For open pit mining, some engineers suggested placing a primed cartridge every 25 feet in a column of explosives for simultaneous detonation of the entire charge. They also suggested lowering the dynamite cartridges either on the end of a rope or by a set of specially designed tongs. In some cases larger cartridges were chopped up and dumped in. Most mining engineers warned against letting large-diameter cartridges free-fall. It also caused a cushion of air to be trapped underneath them, thereby lessening the effect of the explosion[20]. Permitting cartridges to free-fall also risked knocking chunks of the hole's side walls loose, interfering with the consistency of the charge and possibly plugging the hole.

Figure 16. Many open pit mining companies used blasting powder as their primary explosive because it was inexpensive and it performed well. A blasting crew stands amid numerous empty kegs around a drill-hole at an open pit mine. The laborer at right is holding a spool of Cordeau blasting cord which is lashed to a large-diameter dynamite cartridge held by the worker at center. The dynamite cartridge was used to set off the column of blasting powder. (Hercules Inc.; Source: Explosives Engineer May 1926, p 158).

Once the blasting crew loaded the charges, they dumped stemming material into the hole and tamped it down with a large tamping block suspended either from the mast of a churn drill or from a large tripod. Rarely did the crew tamp the column of explosives, they only tamped down the stemming. There was an exception when large-diameter cartridges were not available and the crew had to slice up small, conventional cartridges. The crew usually cut the cartridges in half and pitched them into the drill-hole, creating dead space which necessitated tamping.

Some pit bosses and engineers working in soft rock found an unconventional loading practice known as deck loading to be effective and economical. Deck loading was a means of spreading a small explosive charge evenly throughout the drill-hole by staggering wooden cells in between dynamite cartridges[21]. This practice spread the blast outward, and maximized the breakage while minimizing the use of explosives, there-by saving money. This practice was

very similar to loading dummy cartridges between live ones in underground mines, but in the case of open pit mines, quarry crews used primacord to link the staggered cartridges, avoiding misfires.

Another unconventional loading scheme designed to minimize the cost of explosives involved filling half of the holes in a pattern with less-expensive blasting powder, and the other half with high explosives in a staggered, checkerboard fashion[22]. The reasoning behind this practice was that the high explosives shattered and fragmented the rock while the blasting powder, which had a slower, heaving explosion, picked up the mass and moved it. This technique proved efficient in moderately strong but friable rock.

Figure 17. Two members of a blasting crew lower a large-diameter dynamite cartridge lashed to Cordeau into a churn drill hole during the 1940s. Rarely did blasting crews permit such cartridges to freefall down the hole. (Hercules Inc.; Source: Explosives Engineer *Sept.-Oct. 1944, p 215).*

In some open pit mines, engineers successfully mixed explosive types within drill-holes, stratifying low-velocity bulk explosives with high-velocity dynamites to mirror the rock strata encountered in the pit. The height of the columns of low-velocity explosives matched the softer rock strata, and high-velocity dynamites were situated in hard rock layers. In some cases, engineers found that loading a quick-acting gelatin at the top of a drill-hole and extra-dynamite at the bottom to be very effective for blasting hard rock. When the charge detonated, the gelatin blasted away the hard overburden and provided a zone of weakness for extra-dynamite to blast the underlying rock. In other cases the reverse held true.

Because drilling and blasting in open pit mines was such a specialized operation and had to be adapted to the geology of specific mines, there was little standardization regarding the best quantity of explosives to load. Some open pit mines in hard, igneous rock formations, such as the Utah Copper Company's mines in Bingham Canyon, Utah loaded their churn drill holes nearly to the collar, while in soft rock, crews loaded holes one-half to two-thirds full[23].

Figure 18. The photo captures a blasting crew in the mid-1920s engaged in the process of loading a churn drill hole pattern with dynamite. Several laborers are tamping holes shut with poles and another worker is laying lines of Cordeau blasting cord, which will be used to link the charges. (Hercules Inc.; Source: Explosives Engineer *Aug. 1927, p 282).*

Shooting a Round In the Pits

Once pit crews had loaded, primed, and sealed all of the drill-holes, the round of charges were ready for shooting. Usually, the final preparations for the blast fell into the hands of the shot firer, also known in the mining industry as the powder monkey and blaster. The nature of blasting in open pit mines was complex, involving knowledge of the spectrum of commercial explosives, firing orders, blasting tools and accessories, and how to systematically link numerous high-volume charges. Such complexity often required attention from an explosives specialist, which was different from underground mining where blasting was ore straighforward and fell to the responsibility of small crews.

If the charges in a pit were to be fired with safety fuse and prima-cord, the shot firer examined all of the primacord fuses projecting from the mouths of the sealed drill-

Figure 19. In rare cases well-financed open pit mining companies supplied their blasting crews with special tongs for lowering large-diameter dynamite cartridges into drill-holes, such as the model illustrated. (Source: E.I. DuPont de Nemours & Co. 1920, p 64).

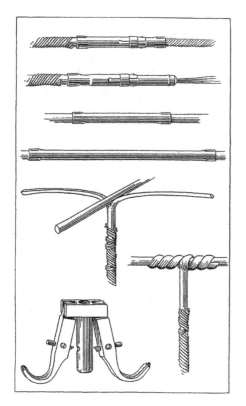

Figure 20. Special supplies were necessary when crews blasted with Cordeau. The illustration shows four means of splicing the ends of two Cordeau lines together at the top, a method of tying individual charges to a trunk line at center, and a special tool for safely splitting Cordeau lengthwise at bottom left. (Source: E.I. DuPont de Nemours & Co. 1934, p103).

holes, and he and an assistant linked them to primacord trunk lines laid across the area to be blasted. The shot firer inspected the wires leading to the capped drill-holes and connected them to reels of blasting wire he paid out to the blasting station. Usually the shot firer used a blasting machine with a 50 to 100 cap capacity, and achieved the desired firing order with delay action caps assigned to pit crews working specific holes. For exceedingly large blasts and complex firing orders, the shot firer may have prepared several independent circuits, which he shot separately.

One of the difficulties of shooting a round in an open pit mine was the logistics of making sure the mine was clear of people and equipment. Blasting in open pit mines was done on a large scale, which meant that a broad area had to be inspected, cleared, and double-checked. When satisfied, the pit crew sealed off the area while the shot firer inspected the holes and made the necessary preparations. Organized mining companies sounded 30 minute, 15 minute, 5, and 1 minute warnings with steam whistles in the early days, and air-horns in the later years. Once the shot firer finished preparations, he initiated the final countdown, shouted the traditional "fire in the hole", and shot the round. Like underground miners, the blasting crew and shot firer listened to and watched the blast to ensure all of the charges were accounted for. The signs of a proper blast consisted of a sequence of pops, followed by thunderous rumblings, the rise of enormous blisters of fractured ground, and finally a pall of smoke and dust. Once the shot successfully fired, the "all clear" whistle sounded and the pit crews returned to work. Drillers returned to their machines to bore the next round of holes, power shovels were brought forth to clear away the shot rock, and another thousand tons of ore was shipped off to be milled.

Figure 21. Open pit mine workers had often considered blasts exciting because the large quantity of explosives and numerous drill-holes contributed to spectacular shots. The blast illustrated is a relatively small shot in an open pit mine. Note the toe shots at the foot of the area blasted, and note the steam shovel and loaded ore hopper cars at right. (Hercules Inc.; Source: Explosives Engineer *March 1925, p 110).*

"Coyotyeing"

During the 1910s engineers and geologists discovered that open pit mining methods lent themselves well to exploiting non-metallic mineral resources. A problem encountered in some of these pits, particularly in friable sedimentary formations where mining companies sought road ballast or argillacious materials for making cement and clay, was that the ground was not cohesive. The soft, friable rock was too resistant to excavate with power shovels, yet it was too soft for conventional open pit mining methods; rockdrill holes and churn drill holes collapsed, and bench walls slumped to form dangerously high and steep banks.

One solution to the problems associated with mining soft rock was a system known as coyoteing, tunnel blasting, and gopher-holing, and it combined underground mining technology with open pit mining's emphasis on the use of massive explosive charges. To prepare for a coyote blast, miners or skilled laborers drove a tunnel into the toe of the mass that was to be blasted, and sent drifts ninety degrees off the tunnel. Workers packed the drifts and the tunnel to the ceilings with explosives, and when detonated, the huge explosion blasted out the toe of the rock mass, causing the overburden to

162   Chapter Four

slough down for easy removal. Coyoteing proved very effective, and engineers subsequently applied the techniques for opening large road cuts, and working ballast and rip-rap quarries[24].

The slang names "coyote tunnel" and "gopher hole" were only slight exaggerations of the actual forms taken by the confined underground workings. Tunnels were typically four to five feet high, three to four feet wide, and the length was usually 2/3 the height of the hillside to be blasted. Miners or laborers excavated material with picks, they occasionally used drilling and blasting methods borrowed from coal mining, and they hauled waste rock out in wheel barrows. Engineers often found driving one set of lateral drifts 50 feet in from the tunnel mouth sufficient to contain enough explosives to effectively rip out the toe of the rock formation, although some blasts required another set of drifts 100 feet farther in[25].

The explosives engineers and contractors preferred for coyoteing were of the inexpensive, slow, low-velocity variety, and the two most favored were blasting powder and railroad powder. They occasionally used ammonium nitrate for blasting relatively harder rock, but it was more expensive.

Loading the explosives was a quick and simple process. Workers placed one case of a high-velocity explosive, such as gelatin, in each drift, and one at the end of the main tunnel, and all of the boxes were primed with non-delay electric caps. On and around the primed cases of dynamite laborers stacked unopened kegs of blasting powder or boxes of railroad powder, in some instances numbering in the tens of thousands. Occasionally, the crews poured

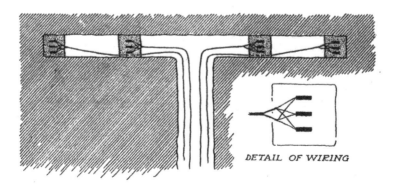

DETAIL OF WIRING

Figure 22. The line drawing is a plan view of a typical coyote blast-hole loaded with explosives. Mining engineers recommended that coyote holes be T-shaped, loaded with blasting powder, and primed with boxes of dynamite and electric blasting caps. The top of the T was parallel to the toe of the hill to be blasted. (Source: E.I. DuPont de Nemours & Co. 1932, p 100).

the blasting powder or railroad powder into huge wood bins set in the floors of the drifts to ensure complete combustion[26].

When all the explosives had been loaded, the entire area was cleared of crew and equipment, and the shot firer inspected the blasting circuit. Satisfied, he shouted "fire in the hole!" and slammed the blasting machine's handle down, which detonated the primed boxes of dynamite buried under the kegs and boxes of powder.The heat and concussion of the exploding dynamite set off the rest of the charge, which blasted away a huge chunk of ground in quite a spectacular show. With no support, the overlying mass of friable, loosely consolidated rock slumped down with a great rumble, sending up a dust cloud as dense as a London fog. Weeks of backbreaking, dangerous excavating and the ant-like procession for loading the thousands of boxes and kegs of explosives had abruptly ended in the matter of minutes.

CHAPTER FIVE

QUARRYING AND BLASTING

Humankind has been quarrying stone for construction for thousands of years. Before the practice of blasting became widespread, quarrying relied on labor-intensive methods which were painfully slow. The introduction of blasting in the seventeenth century revolutionized quarrying as it did mining. Like miners, quarrymen used the new practices of blasting sporadically during the 1700s and with increasing frequency during the early 1800s, until William Bickford invented safety fuse in 1831. Safety fuse revolutionized quarrying from both a humanitarian standpoint and a production standpoint.

Figure 1. This lithograph, published in 1888, depicts a typical small dimension-stone quarry. While specific machines changed in subsequent decades, the basic process remained the same. At left is a portable locomotive boiler supplying steam for piston rock drills and a single cylinder steam hoist. The hoist operator is manipulating the brake as the machine lowers a stone block onto a rail car, with the assistance of a swing boom derrick. Such derricks, semi-portable, were a necessity for moving stone blocks and machines in quarries. The drill operators in the lithograph center are preparing a line of blast-holes for breaking loose more blocks of stone. (Source: Ingersoll Rock Drill Co., 1888, p 19.)

Quarrymen, like miners, found that the use of safety fuse reduced the number of instances in which they blew themselves up, and they found the concept of shooting many holes in a firing order to be quite efficient.

Although the quarrying industry, as it existed during the nineteenth and twentieth centuries, shared with mining a total reliance on drilling and blasting, quarrying would always differ from mining because native rock was cautiously drilled and blasted and dressed into a product, whereas mining discarded it as waste. Quarrying dimension-stone was a very touchy process; fracturing and shattering the desired rock into useless chunks proved to be a mistake not easily avoided. Most stone companies worked their quarries in benches, but some quarry pits had to be worked in shapes and sizes governed by geologic conditions[1]. Generally cohesive, consistent, and

Figure 2. By the mid-1920s, when this photo was taken, quarrying machinery had changed since the late 1880s, but basic quarry methods stayed the same. Quarry workers removed blocks of stone in benches cut from bedrock. The operation in the photo, well-financed, employs two steam-driven channeling machines on short rail lines outlining huge blocks of stone at center, which quarry crews will blast loose. The company erected three swing boom derricks for removing cut stone blocks, the channeling machines, rail, and rockdrills. Most stone quarries were not as large as the operation illustrated. (Courtesy Hercules Inc.; Source: Explosives Engineer *Aug. 1929, p 309.)*

massive types of rock such as sandstone, limestone, marble, quartzite, and granite were conducive to quarrying, and slate was favored because it could be had in large sheets. Quarrymen used explosives to break bedrock into regular blocks which were blasted and cut into smaller ones that stone cutters dressed into dimension-stones. Like underground and open pit mining, the blasting process in dimension-stone quarries included drilling blast-holes, choosing the most effective explosive, priming and loading it, and shooting the round. Exact quarrying methods had to be adapted to the specific rock body being worked, yet quarrymen and bosses were able to devise generalized blasting processes.

By the mid-nineteenth century, in terms of technology, dimension-stone quarrying differed little from mining, because both industries were totally reliant on the use of explosives to maintain profitable levels of production. Both industries applied similar blasting methods, and drilling holes for loading explosives was the beginning of the process in quarries, as it was in mines. The drilling technologies quarries relied on paralleled those of underground mining, but their evolution lagged up to a decade behind.

Before dimension-stone could be produced, the inconsistent, fractured bedrock had to be stripped away from the sound, stratified rock the quarry company sought, and it was usually accomplished through drilling and blasting. Once the sound rock lay exposed, quarrymen conducted further drilling to define and part stone blocks. Quarrymen originally drilled blast-holes by hand with single, double, and triple jacking methods, and most of them favored triple jacking because it was quickest and there was ample space for a crew of three to work. Quarry workers often used single jacking methods to drill the rows of small-diameter holes they used to wedge stone blocks loose.

As was typical of most of the minerals industry in North America, labor-intensive methods in stone quarries did not last, and similar to underground mines, well-capitalized quarry companies mechanized faster than poorly financed operations. Shortly after piston rockdrills proved themselves in railroad tunnels and mines, they supplemented the back-breaking hand-drilling on many quarry benches by the 1890s. While piston-drills were used to bore blast-holes in place of triple jacking crews, they were not always used in place of single jackers to bore the holes quarrymen used for wedging stone blocks loose. The quarrying industry's employment of drills on a large scale paved part of the way for open pit mining, because the drills' widespread surface work formed a model which open pit mining companies later adapted. Some of the specialized rockdrill mountings favored for open pit mining, such as the quarry bar and the gadder, were pioneered in dimension-stone quarries. These two types of drill mounts permitted quarrymen to bore neat, straight lines of drill-holes which facilitated blasting uniform stone blocks. And because little time was lost moving the drills, production was expedited.

In terms of rockdrills, technological change in dimension-stone quarries paralleled that of open pit mining. During the 1910s, most quarry companies found that the old tried and true piston-drills outpaced the new hammer-drills when crews drilled down-holes. And, like the open pit mining industry, quarry companies sent their tired piston-drills to the scrap heap during the early 1920s when drill makers finally offered faster hammer-drills. The quarry industry embraced the *sinker* rockdrill, because, although sinkers were no substitute for deep-drilling drifters, quarrymen found them excellent for drilling blast-holes in shallow

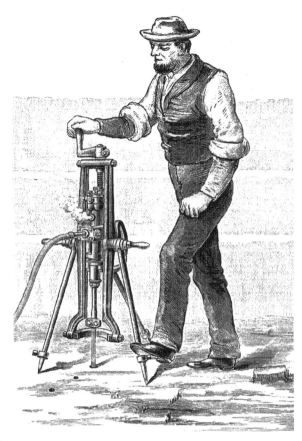

Figure 3. By the mid-1880s rockdrill manufacturers sold piston-drills specifically designed for quarry work, such as the Ingersoll Rock Drill Company's "Little Wonder" drill. Ingersoll's machine is small, relatively light-weight, and permanently fixed onto a tripod mount. (Source: Ingersoll Rock Drill Co., 1888, p 18.)

benches, in large stone blocks that needed to be reduced, and in areas where setting up a drifter-drill on a tripod was difficult.

The *channeler*, developed in the 1890s, was one of the few pieces of machinery explicitly designed for quarrying dimension-stone. As implied by the name, the channeler cut deep channels in living rock, thereby isolating blocks of stone and making them easier to blast out. To cut a channel, the machine hammered a special steam or compressed air-driven gang-bit against the rock as it ever so slowly crept along on a baby-gauge rail carriage. Most of these machines could cut vertical channels up to sixteen feet deep, and some were adjustable for cutting horizontal channels up to seven feet deep[2]. A channeler could have outlined an eight-by-ten foot block of marble, a twenty-by-ten

Figure 4. Like many open pit mines, the dimension-stone quarrying industry did not relinquish their old piston-drills for hammer-drills until the 1920s. This scene, captured in a granite quarry during the early 1920s, typifies medium and small-sized operations. Using piston-drills mounted on tripods, a drilling crew bores a series of closely spaced holes, which will outline large stone blocks when blasted. The worker at far right is running a drill mounted on a quarry bar. Each driller has a tub of water for cleaning rock cuttings out of the hole, possibly with the assistance of compressed air. The holes appear to be fairly deep, judging from the long drill-steels leaning against the pit walls. (Courtesy Hercules Inc.; Source: Explosives Engineer March 1923, p 24.)

foot block of sandstone, or a six-by-ten foot block of slate during one shift, when in the hands of a capable crew[3]. Engineers designed channelers' gang bits for easy replacement, in part because, like rockdrills, they dulled quickly and required sharpening, and because different bit patterns proved efficient for cutting different rock types. Some rockdrill makers also offered channeling bits compatible with the chucks of piston-drills, which pit crews occasionally used in place of track-mounted channelers.

Loading explosives was an integral step of the blasting process conducted in most quarries in North America. Because the blasting technology used in quarries virtually mirrored that used in underground mines, many of the methods quarrymen exercised in priming and loading charges were the same, but some were adapted specifically to parting stone rather than shattering it. Quarrymen also used a variety of drill-hole patterns to define and blast loose stone blocks, some of which were similar to patterns used in underground stopes, and others served exclusively the needs of quarrying.

Figure 5. Quarry bars ranged from sophisticated mechanized mountings which facilitated quick and easy readjustment of the drill, such as the device illustrated, to sawhorses constructed of heavy pipes. Generally well-financed operations purchased the costly mechanized quarry bars in efforts to expedite production, and quarry workers found them easy to work with. Note the pattern of closely spaced vertical and horizontal drill-holes, used for blasting loose granite blocks. (Source: Rand Drill Co. 1886, p 14.)

Like underground mines, quarries could be found in varying sizes, levels of capitalization, and degrees of technological advancement. Blasting duties in most small quarries were handled by work crews, and they used simple tools and supplies for loading and shooting their rounds of charges. Most large operations had a sufficient income and levels of production to afford to hire shot firers and provide them with expensive tools and supplies. By the 1900s many well-capitalized quarries were following the trend of hiring shot firers to conduct the blasting, to the exclusion of the drillers and other laborers[4]. Regardless of who did the blasting in a stone quarry, there were a number of steps that had to be taken prior to shooting the round of holes completed by the drillers.

Quarrymen, like miners, faced the choice of the best explosive for their needs, and the criteria boiled down primarily to the explosives' performance, and cost secondarily. The performance expected of an explosive consisted of being able to fracture the rock without shattering it, which is why most quarrymen prior to the 1920s preferred blasting powder. Its soft, heaving explosion was more conducive to cracking stone and leaving intact blocks, than most types

of dynamite. Quarrymen also favored railroad powder because its performance was similar to blasting powder, but it had a little more kick which was better in some types of harder rock[5]. The best grain-size of either explosive they chose depended on the type of rock blasted; a slow, course-grained powder was best in softer rock such as sandstone, and in harder rock such as quartzite, granite, and slate, quick, fine-grain powder was best.

Some quarries in hard rock did employ low-percentage dynamites, ammo-nium nitrate, nitro-starch, or gelatins when their quicker explosions were necessary. This became especially true during the late

Figure 6. Rockdrill makers produced gadders as early as the mid-1880s, such as the Ingersoll Rock Drill Company's versatile model show in the illustration. A special mount on the boom of Ingersoll's gadder permitted the driller to bore a series of consistent horizontal or vertical blast-holes, as shown. Because quarry floors tended to be far too rough to permit workers to freely move such a heavy machine on iron wheels, most gadders ran on rails. Note the upright boiler in the background, which supplied steam to power the gadder. (Source: Ingersoll Rock Drill Co. 1888, p 21.)

1910s when explosives makers began offering dynamites specifically geared toward quarrying. More and more quarrymen experimented with these relatively new products and found their performance to be as good as blasting powder in some conditions. By the 1920s most of the major explosives manufacturers offered explosives under brandnames such as *Quarry Gelatin, Quarry Special,* and *Quarry Dynamite.* As the quarry industry progressed

Figure 7. Well financed stone companies working in hard rock such as marble and granite employed channeling machines. Driven by steam or compressed air, channelers operated on the same principles as rockdrills, ramming a special gang bit into bedrock. As the bit hammered away, the machine slowly crept forward on a rail line, creating a deep channel in rock as it moved. The lithograph shows an Ingersoll Rock Drill Company channeler circa mid-1880s. Note the gang bit at left and the gearing for the motive mechanism on the main frame. (Source: Ingersoll Rock Drill Co. 1888, p 22.)

through the 1920s, blasting powder was no longer the undisputed favorite explosive poured into quarry drill-holes. Specifically formulated and low-percentage dynamites saw increased application.

Like underground mines, successful blasting in quarries was totally dependent on drill-hole patterns and loading explosives without error. But quarrying differed in that the patterns were designed to create planes of weakness to facilitate controlled fracturing, and in some instances specialized loading systems had to be used to overcome the limitations associated with blast-holes drilled by hand. The specific hole pattern chosen by quarrymen was a function of the drilling technology they used, and the type of rock being quarried.

Prior to the widespread acceptance of machine drills in the quarrying industry in the 1910s, crews usually bored blast-holes by hand. One of the biggest problems with hand-drilled holes was that they tended to be triangular in

Figure 8. Large channelers were self-contained and featured their own upright steam boilers for power. While these machines were expensive and cumbersome to move around the quarry, they eliminated the need for a costly compressed air plant. Well-financed operations employed self-contained steam channelers for work in hard stone into the 1920s. The lithograph depicts a scene at a large, mechanized stone quarry during the mid-1880s. Note the replacement gang bit at lower left. (Source: Ingersoll Rock Drill Co. 1888, p 23.)

shape, and when an explosive charge was shot in such a hole, fractures tended to radiate from each corner of the triangle instead of in the square pattern desired by quarrymen[6]. To overcome this problem, quarrymen devised several systems combining special methods of loading powder and arranging hole patterns.

The first method, known as *Lewising* and as *broaching*, was used in Maine granite quarries at least as early as the mid-nineteenth century[7]. This method required that drillers bore several very closely spaced holes in a series, and break out the rock separating them. With this method, quarrymen created rough channels which defined a block of stone. The person conducting the blasting loaded each channel with powder-filled tin canisters suspended in a matrix of sand. The sand cushioned the blast, and the channel's length, depth, and orientation caused its force to fracture the rock along desired planes.

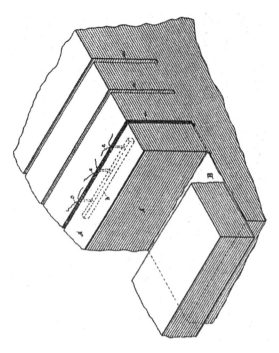

Figure 9. The U.S. Patent block diagram depicts a variation of Lewising, in which workers broke loose large stone blocks by cutting a channel, filling most of the void with sand, pouring blasting powder over the fill, and packing tamping material over the powder. The explosion of the small, long charge of powder was enough to crack loose the stone block while leaving it intact. Lewis D. Conner of Oberlin, Ohio, patented this process in 1890, possibly coining the name "Lewising." (Source: U.S. Patent Gazette, March 25, 1890, p 1604.)

The *Portland Canister System*, developed in granite quarries near Portland, Maine, was also used to overcome problems associated with triangular drill-holes[8]. Quarrymen carrying out this method defined stone blocks with a crosshatching of holes. The defined blocks ranged from three-by-three feet in area to twelve-by-twelve feet, and each hole was approximately three inches in diameter. The quarry crew loaded the holes with canisters of blasting powder suspended in a matrix of sand, and they shot the round. Once the charges fired, quarrymen selected wide fractures along the desired planes and hammered in a series of steel wedges, further opening the fractures. In some cases they poured a little blasting powder straight into the hole and shot it to help spread the cracks[9]. By wedging, blasting, and prying, eventually the quarry workers won fairly clean blocks of stone.

The third method, known as the *Knox System*, required a series of drill-holes, each reamed into an ovoid shape in footprint, and spaced four-to-twelve feet apart, depending on the hardness of the rock[10]. A quarryman loaded the holes with bulk blasting or railroad powder and shot them, and the stone broke along the planes of weakness created by the pattern. One factor common to all of the above patterns was that setting off all of the holes associated with a single stone block simultaneously, rather than in an order, promoted even fracturing.

A fourth and commonly used pattern was a variation of the Knox System, and it consisted of several series of closely spaced drill-holes defining multiple blocks. Such an arrangement of holes, when fired simultaneously, cracked loose large blocks of rock which could then be reduced into dimension-stones either with mechanical means, or by blasting.

A fifth method of generating stone blocks involved drilling a row of vertical blast-holes and shooting them to create a vertical plane of fracture, and reducing the defined block with *plugs and feathers* hammered into a row of horizontal holes. Usually the machine drill-holes were larger in diameter and depth than the single jacked horizontal holes. Plugs and feathers consisted of a hardened steel wedge which was hammered between two softer iron shims, and when numerous sets were used simultaneously, the force they exerted cracked the rock. This ancient but effective method was used into the twentieth century for isolating, breaking loose, and reducing stone blocks.

Loading blast-holes made by machine drills, and those associated with the above two patterns, was a quick and simple process. The most common way quarrymen accomplished the task was by pouring blasting or railroad powder straight in, and tamping the hole closed, like the method used by underground miners. When quarrymen used dynamite instead of powder, they made up primers, slit the cartridges so they could expand when tamped them down, inserted the dynamite, and sealed the hole. In cases where the rock was brittle, quarrymen tamped a little paper wadding over the charge to create a space to cushion the initial shock of the powder's explosion, and then they sealed the hole. For the most part, quarry crews filled approximately one-third of each drill-hole with explosives[11].

Like underground mining, quarrymen primed blasting powder with safety fuse, and they inserted blasting caps into dynamite and railroad powder charges. In later years, some quarries successfully used electric squibs. If the blasting conditions were wet, then quarrymen had to make up cartridges to protect bulk blasting powder and railroad powder. Once all of the charges were primed, loaded, and sealed, the crew conducting the blast surveyed the rock bench to ensure all was well. What they should have seen prior to shooting the round was a pattern of clay blobs representing the sealed drill-holes, embedded in bedrock.

The process of shooting a round of blast-holes in quarries mirrored the methods used in underground mines, and as with mining, the process quarrymen exercised depended on the source of ignition or detonation they used. Prior to the 1910s most quarries primed their blasting powder charges with safety fuse and their dynamite with caps and fuse, in the fashion of underground miners.

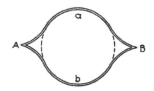

Figure 10. The diagram illustrates a plan view of a knox drill-hole. Workers reamed the hole and created flutes like those in the illustration. When shot, a row of aligned fluted holes directed the force of a blast along a determined plane, cracking loose large blocks of stone. (Source: Gillette 1907, p 197.)

Figure 11. The early lithograph depicts a driller boring one of the most common hole patterns employed by quarry companies for parting stone blocks. The pattern consisted of a series of closely spaced drill-holes, which created a line of fracture when shot with small powder charges. Rockdrills mounted on quarry bars lent themselves well for boring series of holes. The lithograph illustrates fractured vertical faces resultant from blasting such hole patterns. (Source: Rand Drill Co. 1886, p 12.)

Quarrymen lit the safety fuses with rat tails just like their underground brethren, but once explosives manufacturers offered affordable and reliable electric blasting equipment in the 1910s, the better-capitalized quarries embraced that technology. On many quarry benches electric blasting gave better results than the tried and true safety fuse and standard caps, because blasting machines shot electric caps virtually simultaneously, which evenly fractured the rock. In later years some quarries used primacord, which gave results similar to electric blasting. Electric blasting gained popularity through the 1910s until nearly all large quarries used it by the 1920s. Still, throughout the first half of the twentieth century small quarries proliferated, and they continued to used the old-fashioned safety fuse and conventional caps.

After lighting the fuses, the quarry crew, or the *shot firer* employed by large operations, retreated to a place of safety and awaited the impending blast. In the quarries where shots were fired electrically, a small crew connected the caps into a circuit that was wired to a blasting machine. One of the senior crew members slammed the machine's handle down, energizing the circuit. Like seasoned underground drilling and blasting crews, the quarry workers listened for the reports of the charges to ascertain whether they all detonated. Satisfied, a few crew members rose after the smoke cleared to inspect the blast's outcome, followed by stone cutters who began clearing the new blocks away with ropes and derrick hoists for trimming, and drillers returned to their oily machines.

ENDNOTES

Chapter 1: Explosives Used for Mining

1. **Davis, Tenney, 1990,** p 48.
Gillette, Halbert, 1878, P 108.
International Correspondence Schools, 1907, A. 36: 7.

2. **Andre, George, 1878,** p 98.
Cabinet of Miscellaneous Info.
Young, George, 1923, p 103, 245.

3. **Davis, Tenney, 1990,** p 2.
Hercules Powder Co., 1918, p 13.

4. Cabinet of Miscellaneous Info.

5. **Chellson, H. C., 1936.**
VanGelder, Arthur and Schlatter, Hugo, 1972, p 11.

6. **Chellson, H. C., 1936**
Young, Otis Jr., 1982, p 187.

7. **Marvin, Theodore, Editor, 1936.**

8. **VanGelder, Arthur and Schlatter, Hugo, 1972,** p 27.

9. **Davis, Tenney, 1990,** p 49.
Hercules Powder Co., 1918, p 64.

10. **Davis, Tenney, 1990,** p 3.

11. Cabinet of Miscellaneous Info.
International Textbook Company, 1899, A. 41: 67.

12. **Andre, George, 1878,** p 102.
Gardner, Edward, Editor, 1927, p 4.
Warwick, A. M., 1902
Young, George, 1923, p 103, 245.

13. **Hercules Powder Co., 1918,** p 17.

14. **Martin, Andy, 1991,** p 70.

15. **California Cap Company, 1910,** p 10, 20, 56.

16. Table 2 was compiled from archaeological data collected in historic Western mines, and from Andy Martin's *Blasting Cap Tin Catalog* (1991).

17. **Davis, Tenney, 1990,** p 196.
Guttman, Oscar, 1906, p 9.
McAdam and Westwater, 1958, p 3.

18. **Davis, Tenney, 1990,** p 206-207.

19. **Andre, George, 1878,** p 88.
Davis, Tenney, 1990, p 207.

20. **VanGelder, Arthur and Schlatter, Hugo, 1972,** p 321.

21. Ibid, p 321.

22. Ibid, p 323.

23. Ibid, p 324.

24. Ibid, p 325.

25. **U.S. Blasting Oil Co., 1866,** p 5.

26. **Shaffner, T P, 1869,** p 21.

27. **VanGelder, Arthur and Schlatter, Hugo, 1972,** p 328.

28. **Davis, Tenney, 1990,** p 336.

29. **VanGelder, Arthur and Schlatter, Hugo, 1972,** p 432.

30. Ibid, p 330.

31. Early noteworthy active-base dynamites include Dr. Carl Volney's *Rendrock* with 40% nitroglycerine, 40% saltpeter, 13% alkali-treated sawdust, and 7% paraffin patented in 1873 (U.S. Pat. 157,143); Egbert Judson's *Giant Powder No. 2* made with 40% nitroglycerine, 40% saltpeter, 6% sulfur, 6% rosin, and 8% kieselguhr, also patented in 1873 (U.S. Pat. 139,468); Joseph Willard's *White Hercules*, patented in 1874, with an active base of saltpeter, magnesium carbonate, and sugar (U.S. Pat. 157,054), and Robert Warren's *Vulcan Powder* patented in the mid 1870s.

32. Archaeological evidence at many historic Western mines that were abandoned in the 1880s reflects a transition from dynamite to blasting powder during this decade. Artifact assemblages in specific mines include primarily dynamite boxes and very few blasting powder kegs, some vice-versa, and some mines feature containers for both explosives. In mining districts abandoned between the 1890s and 1910s archaeological records include very few powder kegs and mostly dynamite boxes and blasting cap tins, indicating blasting powder had been superseded by dynamite by this time.

33. **Taylor, W. W., 1931.**

34. **Marvin, Theodore, Editor, 1931.**

35 General Mining News: Colorado, Teller County.

36. **Hubbard, Douglass, 1958,** p 6.

37. **Stark, Jared, 1910.**

38. **Tratman, E. E., 1893.**

39. **Shaffner, T. P., 1869.**

40. **Tratman, E. E., 1893.**

41. **Johnson, W. T., 1903.**

42. **Stark, Jared, 1910.**

Wilson, Eugene, Editor, 1910b.

43. U.S. Patent # 540,141.

44. Mine Inspectors' Reports: File 48441 Vindicator Mine.

45. **McGraw, James, 1901a.**

46. **Wyman, Mark, 1979,** p 132.

47. Burning Empty Dynamite Cases. **Dannenburg, Joe, 1960-1980,** Pt. 1: 40.

48. **Davis, Tenney, 1990,** p 339. **VanGelder, Arthur and Schlatter, Hugo, 1972,** p 339. U.S. Patent # 175,735.

49. **E.I. DuPont de Nemours & Co., 1932,** p 17.

Hercules Powder Co., **1918,** p 36.
Peele, Robert, 1918, p 146.

50. **VanGelder, Arthur and Schlatter, Hugo, 1972,** p 340.

51. Ibid, p 340.

52. The rise in popularity of extra formulas is reflected in the archaeological records at historic mine sites, and in modern artifact collections. Boxes for dynamite extra and gelatin extra predating 1915 are relatively rare while those post-dating 1920 are common.

53. U.S Patent # 312,010.
VanGelder, Arthur and Schlatter, Hugo, 1972, p 341.

54. **VanGelder, Arthur and Schlatter, Hugo, 1972,** p 637-639.

55. U.S. Patent # 183,764.
VanGelder, Arthur and Schlatter, Hugo, 1972, p 338.

56. **Hall, Clarence, 1911.**

57. **Nobel's Explosives Co., Ltd., 1887.**
VanGelder, Arthur and Schlatter, Hugo, 1972, p 343.

58. U.S. Patent # 839,450.

59. **Cohen, Stan, 1984,** p 106.

60. The prices of dynamite and blasting powder were determined from historical explosives and mine supplies sales receipts, and from the source listed below. Prices for explosives varied according to numerous variables such as vagaries within the explosives industry, the mode of transportation (wagon or railroad), the quantity purchased, and the distance from explosives makers and their major distribution offices to the purchaser.

Young, George, 1923, p 104.

Chapter 2: One Round In and One Round Out: The Technology of Mining and Blasting

1. **Andre, George, 1878,** p 129.
Sporr, Ray, Mining Engineer and Gold Miner 1930s - 1950s, 1993.

2. Ibid.

3. Ibid.
Watson, Tom: Gold Miner 1930s, 1988.

4. **Gillette, Halbert P., 1907,** p 15.
Hoover, Herbert C., 1909, p 150.
International Correspondence Schools, 1907, p 13.
Peele, Robert, 1918. p 184.
Young, George, 1946, p 87.

5. A reveiw of Colorado State Mine Inspector's Reports from the Cripple Creek Mining District recorded during the 1930s revealed numerous examples of hand-drilling. The source listed below indicated that miners drilled by hand during the Great Depression. Sporr, Ray, Mining Engineer and Gold Miner 1930s - 1950s, 1993.

6. **Stack, Barbara, 1982,** p 26.
Ingersoll-Rand Drill Company, 1939, p 3.
Young, George, 1946, p 111.

7. **Johnson, Walter, Mining Engineer 1930s - 1960s, 1993.**
Peele, Robert, 1918, p 185.

8. Ibid, p213.

9. **Wyman, Mark, 1979,** p 90.
A survey of historic mine sites in the West reflects the fact that few mine plants predating 1890 featured air compressors and rockdrills, while they were more common among large mines post-dating 1890.

10. **Cleveland Rock Drill Co., 1931,** p 5.
Ingersoll-Rand Drill Company, 1939, p 3.
Stack, Barbara, 1982, p 30.

11. **Ingersoll-Rand Drill Company, 1938,** p 1.

12. **Peele, Robert, 1918,** p 187.

13. **Johnson, Walter, Mining Engineer 1930s - 1960s, 1993**

14. Ibid.

15. **Young, George, 1946,** p 111.

16. **International Correspondence Schools, 1907,** A. 36: 67.
Peele, Robert, 1918, p 221.
Young, George, 1923, p 127-130.

17. **Peele, Robert, 1918,** p 221.

18. **International Correspondence Schools, 1907,** A. 36: 65-68.
Peele, Robert, 1918, p 221.
Young, George, 1923, p 127-130.

19. **International Textbook Company, 1899,** A. 41: 78.
International   Correspondence Schools, 1907, A. 36: 65-66.
Peele, Robert, 1918, p 258.
Young, George, 1923, p 127-130.

20. **Young, George, 1946,** p 142.

21. Ibid, p 155.

22. **Hoover, Herbert C., 1909,** p 95-96.
Peele, Robert, 1918, p 486.
Young, George, 1946, p 554.

23. **Drake, Raymond, 1983,** p 36.

24. **Donovan, Hedley, 1976,** p 64.
Peele, Robert, 1918, p 531.

25. **Young, George, 1946,** p 157.

26. **Cohen, Stan, 1984,** p 11.

27. Ibid, p 11.

28. **Wilson, Eugene, Editor, 1910a**

29. **Cohen, Stan, 1984,** p 11.
Peele, Robert, 1918, p 730.

30. **Stack, Barbara, 1982,** p 504.

31. **Stack, Barbara, 1982,** p 502.

32. **Munroe, Charles, and Hall, Clarence, 1909,** p 36.

33. **E.I. DuPont de Nemours & Co., 1917a,** p 34.

34. **Munroe, Charles, and Hall, Clarence, 1909,** p 36.

35. **Bragg, Melody, 1990.**
Polly, Philip, Tungsten and Coal Miner, 1930s - 1940s, 1993

36. **Koch, Karl, Coal Miner 1920s, 1994.**

37. **Wilson, Eugene, Editor, 1910a.**

38. **Stoek, Henry, 1906b.**

39. **Andre, George, 1878,** p 109.
Munroe, Charles, and Hall, Clarence, 1909, p 37.
Peele, Robert, 1918, p 171.
Stoek, Henry, 1906a

40. **Andre, George, 1878,** p 136.
E.I. DuPont de Nemours & Co., 1917a, p 30.
Munroe, Charles, and Hall, Clarence, 1909, p 41.
Peele, Robert, 1918, p 168.
Stark, Jared, 1910.

41. **Howell, S P, 1931,** p 16-19.
Koch, Karl, Coal Miner 1920s, 1994.
Sporr, Ray, Mining Engineer and Gold Miner 1930s - 1950s, 1993.
Stark, Jared, 1910.

42. **Howell, S. P., 1931,** p 16.

43. **Colliery Engineering Company, 1905,** p 331.
Foster, Thomas, Editor, 1896.

44. **Young, George, 1946,** p 162.

45. **Stoek, Henry, 1906a.**

46. **Muroe, Charles and Hall, Clarence, 1909,** p44.

47. **Stoek, Henry, 1906a.**

48. **Peele, Robert, 1918,** p 171.

49. **Gonzolez, Frank, Coal Miner 1930s - 1950s, 1993.**
Polly, Philip, Tungsten and Coal Miner, 1930s - 1940s, 1993.

50. **E.I. DuPont de Nemours & Co., 1917a,** p 35.

51. **E.I. DuPont de Nemours & Co., 1917a,** p 35.
Sporr, Ray, Mining Engineer and Gold Miner 1930s - 1950s, 1993.

52. **Polly, Philip, Tungsten and Coal Miner, 1930s - 1940s, 1993.**

Sporr, Ray, Mining Engineer and Gold Miner 1930s - 1950s, 1993.

53. **E.I. DuPont de Nemours & Co., 1917a,** p 19.
Explosives and Miscellaneous Investigations.

54. **E.I. DuPont de Nemours & Co., 1917a,** p 19.
International Textbook Company, 1899, A. 41: 87.
Munroe, Charles, and Hall, Clarence, 1909, p 41.

55. **Koch, Karl, Coal Miner 1920s, 1994.**
Sackett, Earl, 1977.

56. **International Correspondence Schools, 1907,** A. 36: 33.
Munroe, Charles, and Hall, Clarence, 1909, p 30.
Peele, Robert, 1918, p 172.

57. U.S. Patent # 148,674.

58. **Munroe, Charles, and Hall, Clarence, 1909,** p 30.

59. **E.I. DuPont de Nemours & Co., 1917a,** p 25.
Peele, Robert, 1918, p 171.
U.S. Patent # 203,097.
Young, George, 1923, p 112.

60. **Munroe, Charles, and Hall, Clarence, 1909,** p 41.

61. **Cohen, Stan, 1984,** p 106.

62. **Andre, George, 1878,** p 133.
E.I. DuPont de Nemours & Co., 1932, p 35.
International Correspondence Schools, 1907, A. 36: 33.
International Textbook Company, 1899, A. 41: 86.
Peele, Robert, 1918, p 172.

63. **E.I. DuPont de Nemours & Co., 1917a,** p 21.

64. **E.I. DuPont de Nemours & Co., 1932,** p 11.

65. **Stark, Jared, 1910.**

66. **Peele, Robert, 1918,** p 175.

67. **Banks, George, 1929.**

68. **Gonzoles, Frank, Coal Miner 1930s - 1950s, 1993.**
Polly, Philip, Tungsten and Coal Miner, 1930s - 1940s, 1993.
Sporr, Ray, Mining Engineer and Gold Miner 1930s - 1950s, 1993.

69. The prices for dynamite and blasting powder were derived from a survey of historic explosives and mine supply receipts from a variety of mining districts, from trade advertisements, from trade catalogs, and from the sources listed below.
Peele, Robert, 1918, p 1284 -1309.
Young, George, 1923, p 104.

70. **VanGelder, Arthur and Schlatter, Hugo, 1972,** p 511.

71. **Young, George, 1923,** p 104.

72. **Peele, Robert, 1918,** p 226.
Young, George, 1946, p 127.

73. Trends pertaining to miners' choice of dynamite percentage strengths is reflected in archaeological records at historic mines throughout the West, and in the assemblages of boxes in extant artifact collections. The vast majority of dynamite boxes encountered at historic mines and in collections are marked *40%*, while boxes once containing *60%*, *30%*, and *20%* are relatively rare. Boxes for 60% dynamite have been encountered in mines in very hard rock such as young granite and some metamorphic rock. Where rock is soft, low-percentage-strength boxes have been encountered. In addition to archaeological evidence, the following sources also have documented trends regarding choice of strength:
Polly, Philip, Tungsten and Coal Miner, 1930s - 1940s, 1993.
Sporr, Ray, <u>Mining Engineer and Gold Miner 1930s - 1950s</u>, 1993.
Watson, Tom: Gold Miner 1930s, 1988.

74. **Anderson, A E, 1912.**

75. **Young, George, 1946,** p 174.

76. **Peele, Robert, 1918,** p 259.

77. **International Correspondence Schools, 1907,** A. 36: 41.
International Textbook Company, 1899, A. 41: 63.
Munroe, Charles, and Hall, Clarence, 1909, p 44.
Peele, Robert, 1918, p 259.

78. **Johnson, Walter, Mining Engineer 1930s - 1960s, 1993.**
Polly, Philip, Tungsten and Coal Miner, 1930s - 1940s, 1993.
Sporr, Ray, Mining Engineer and Gold Miner 1930s - 1950s, 1993.
Watson, Tom: Gold Miner 1930s, 1988.

79. The Author observed blasting caps with teeth marks and plier marks in Western mines. Pliers have also been encountered at underground priming stations in historic mines.

80. **Colliery Engineering Company, 1905,** p 332.
E.I. DuPont de Nemours & Co., 1932, p 59.
Gonzolez, Frank, Coal Miner 1930s - 1950s, 1993.
Peele, Robert, 1918, p 168.
Polly, Philip, Tungsten and Coal Miner, 1930s - 1940s, 1993.
Sporr, Ray, Mining Engineer and Gold Miner 1930s - 1950s, 1993.
Watson, Tom, Gold Miner 1930s, 1988.

81. **Twitty, Eric, 1997**

82. Ibid, p 17.

83. Ibid, p 20.

84. Ibid, p 20.

85. **Colliery Engineering Company, 1905,** p 330.
International Correspondence Schools, 1907, A. 36: 48.
Polly, Philip, Tungsten and Coal Miner, 1930s - 1940s, 1993.
Sporr, Ray, Mining Engineer and Gold Miner 1930s - 1950s, 1993.

86. **Watson, Tom: Gold Miner 1930s, 1988.**

87. Old Powder Said to be Cause of Accident.

88. **Anderson, A. E., 1912.**
Gardner, Edward; Howell, S. P.; Jones, G. W., 1927.
Johnson, Walter, Mining Engineer 1930s - 1960s, 1993.
Polly, Philip, Tungsten and Coal Miner, 1930s - 1940s, 1993.
Sporr, Ray, Mining Engineer and Gold Miner 1930s - 1950s, 1993.

89. **Anderson, A. E., 1912.**
Peele, Robert, 1918, p 226.

90. **Allen, A. W., Editor, 1930.**
Young, George, 1946, p 142.

91. **Hall, Clarence, 1911,** p 14.

92. Evidence of spent wire pull igniters remaining today is almost totally absent at Western hardrock mine sites, indicating they were rarely used.

Chapter 3: Mines, Miners, and Explosives

1. Mining Summary: Colorado, Teller County.

2. **Institute of Makers of Explosives, 1935,** p 11.

3. **E.I. DuPont de Nemours & Co., 1917a,** p 15.
E.I. DuPont de Nemours & Co., 1934, p 33.
Institute of Makers of Explosives, 1935, p 11.
Munroe, Charles, and Hall, Clarence, 1909, p 58.
Peele, Robert, 1918, p 155.
Young, George, 1923, p 110.

4. **Stark, Jared, 1910.**
Magazines such as the one documented by Stark, as well as those noted in the previous text, can still be encountered at historic Western mines. Archaeological evidence at historic mines indicates that well financed operations had anything approaching what could have been called a magazine, and only productive, progressive mines had storage facilities that were properly built.

5. **Sporr, Ray,** Mining Engineer and Gold Miner 1930s - 1950s, **1993.**

6. **Wyman, Mark, 1979,** p 132.

7. **Stark, Jared, 1910.**

8. **Johnson, Walter,** Mining Engineer 1930s - 1960s, **1993.**
Polly, Philip, Tungsten and Coal Miner, 1930s - 1940s, 1993.
Sporr, Ray, Mining Engineer and Gold Miner 1930s - 1950s, 1993.
Watson, Tom: Gold Miner 1930s, 1988.
Stark, Jared, 1910.
Archaeological evidence existing today in historic mines is consistent with the documented behavior of

miners in magazines. Today it is common to find historic safety fuse wrappers, lengths of safety fuse, blasting caps, their tins, wrappers and liners, and dynamite box parts together the same in historic underground and surface magazines.

9. **E.I. DuPont de Nemours & Co., 1917a,** p 13.
Institute of Makers of Explosives, 1935, p 8.
Munroe, Charles, and Hall, Clarence, 1909, p 27.

10. **E.I. DuPont de Nemours & Co., 1932,** p 26.

11. **Howell, S. P., 1931,** p 29.

12. **Howell, S. P., 1931,** p 26.

13. **Koch, Karl, Coal Miner 1920s, 1994.**

14. Blasting powder manufacturers used two types of materials to make their kegs. Explosives makers originally packed powder into wood kegs from a time predating the scope of this work until steel had replaced wood by the early 1900s. Totally hand-made kegs, identifiable by a multi-faceted body, each stave presenting an almost flat face, predated the 1850s. Kegs made after usually featured concentric rings around the body which were created by a stave-dressing machine common in cooper shops post-dating the 1840s (Howard, 1979b). Kegs that utilized cut nails to fasten down the hoops suggests a date earlier than the late 1880s, and wire nails came afterward.

Beset by a number of problems such as leaking, breakage, and loosening, wood kegs proved to be far from ideal for storing and shipping blasting powder. Also, because wood kegs were hand-made by a variety of coopers, they were rarely uniform in size. Aware of these problems, Wilmington, Delaware, tinsmith James Wilson, who manufactured small gunpowder tins for DuPont, experimented with a number of alternatives. His solution, developed and patented in 1857, was the first iron powder keg produced in North America; however, its soldered construction rendered it too expensive to make in the volume needed by the powder industry (U.S. Patent # 16,944).

To overcome high production costs, Wilson's tin foundry, Wilson, Green & Wilson patented yet another iron keg in 1859 that was cost-effective to produce in volume. The secret to the new keg lay in its construction: instead of being soldered together, its side-seam was rolled, and its ends were crimped onto the body by a special set of machines, which the firm had also patented. Wilson's iron keg consequently became the industry-standard continent-wide, and through the latter half of the nineteenth century they began to phase out wooden types. By 1900 few if any powder companies manufactured wood kegs.

Powder makers usually charged a deposit for their kegs to provide miners with incentive to return them for refill, but Western mining districts were frequently so remote that the cost of the deposit was not worth freighting the empties out. In response, the powder companies devised the *cleat*

top keg with the intent of producing a cheap, disposable container. This type of keg was introduced to mining districts in the West around 1886, it gained popularity during the 1890s, and it had phased out most types of reusable kegs by 1910. The cleat top was a sheet metal seal crimped to the rim of the bung hole in place of the more expensive threaded zinc bung previously used. In addition, the sheet steel composing the keg was thinner than for the stout refillable models. While the cleat seal could not be closed as effectively as threaded bungs, the new type of keg cost much less to produce, effectively functioning as a disposable container.

15. **Hercules Powder Co., 1883,** p 4. **Colliery Engineering Company, 1905,** p 329.

16. Old Powder Said to be Cause of Accident.

17. **Tratman, E. E., 1893.**

18. Beginning with the first batch of dynamite manufactured in 1868 by the Giant Powder Company and continuing into the 1950s, most explosives makers packed high explosives into wood boxes for shipment. The standard quantity was fifty pounds, but they also offered tewnty-five and ten pound boxes to miners and prospectors engaged in limited work.

The manufacturing techniques powder companies and box makers used to assemble dynamite boxes follow a distinct chronology. From 1868 until shortly after 1900, the most common and cheapest method of assembling the components consisted of nailing the sides onto the end-panels with three to four nails per joint, nailing on the bottom, and finally screwing down the lid when the box had been filled.

During the late 1880s dynamite box construction changed in three regards. First, makers replaced cut nails with wire nails. Second, because wire nails did not split wood as did cut nails, thinner panels were used. Last, manufacturers dropped use of wood screws to fasten down the lid. These nail-together boxes with thinner panels, so thin in some cases the boxes were downright flimsy, became the industry standard between approximately 1890 and 1907.

Around 1904 dynamite boxes began to change yet again. Some of the larger companies, such as the Giant Powder Company Consolidated, the California Powder Works, the Repauno Chemical Company, the Aetna Powder Company, and the Climax Powder Manufacturing Company began to use lock-corner boxes made out of half-inch-thick wood instead of one-quarter-inch-thick wood. By 1907 this lock-corner box became the standard shipping container for dynamite throughout the explosives industry, totally phasing out nail-together boxes (Hopler, 1988:31). By 1914 the Interstate Commerce Commission (ICC) mandated that all boxes used to ship explosives be of this construction, and to demonstrate compliance dynamite makers printed I.C.C.-14 on their boxes. In 1927 the ICC permitted explosives makers to use *cleat end* boxes, which were nailed together with inset lath frames reinforcing the ends (Hopler, 1988:31). Some

dynamite makers favored this type of box. Another significant change occurred in 1928, when the ICC allowed the bottoms and tops of boxes to reduced from one-half-inch to three-eighths-inch thickness (Hopler, 1988:31). The last major change to dynamite box construction was the ICC's permit to allow explosives companies to phase in *fiberboard* boxes beginning in 1931 (Hopler, 1988:31).

19. **Stark, Jared, 1910.**

20. Ibid.

21. **Wyman, Mark, 1979,** p 133.

22. **Rickard, T. A., 1913.**

23. **Smith, B.H., 1916.**

24. **California Cap Company, 1910,** p 21.
Cleveland Rock Drill Co., 1931, p 105.

25. **Taylor, W. W., 1931.**

26. **Anderson, A. E., 1912.**

27. **California Cap Company, 1910,** p 24.
Cleveland Rock Drill Co., 1931, p 105.
Gillette, Halbert, P., 1907, p 135.
Peele, Robert, 1918, p 227.
Young, George, 1946, p 142.

28. **Howell, S. P., 1931,** p 39.

29. **Crampton, Frank, 1982,** p 216.

30. **Cleveland Rock Drill Co., 1931,** p 105.
Gillette, Halbert, P., 1907, p 200.

31. **Howell, S. P., 1931,** p 41.

32. **Wyman, Mark, 1979,** p 106.

33. **Andre, George, 1878,** p 98.
Cabinet of Miscellaneous Info.
Hercules Powder Co., 1918, p 26.

34. **Andre, George, 1878,** p 80.
Cabinet of Miscellaneous Info.
Gardner, Edward, Editor, 1927, p 5.
Warwick, A. M., 1902.

35. **Rutledge, J. J. and Hall, Clarence, 1912,** p 22.

36. **Wyman, Mark, 1979,** p 107.

37. **Hercules Powder Co., 1882.**
Repauno Chemical Co., 1895.
U.S. Blasting Oil Co., 1866.
Wyman, Mark, 1979, p 105.

38. **Gardner, Edward, Editor, 1927,** p 7.
Hercules Powder Co., 1918, p 26.
Need For Ventilation.

39. **Connibear, William, 1924.**

40. **Wyman, Mark, 1979,** p 107.

41. **E.I. DuPont de Nemours & Co., 1920,** p 13.
Peele, Robert, 1918, p 146.

42. **E.I. DuPont de Nemours & Co., 1920,** p 101.

43. **E.I. DuPont de Nemours & Co., 1920,** p 12.
Hercules Powder Co., 1918, p 36.
Peele, Robert, 1918, p 146.

44. **Hunter, Edward, Mining Engineer 1930s - 1990s.**

45. **E.I. DuPont de Nemours & Co., 1952,** p 251.

46. Some of these primitive systems are still evident at abandoned historic mines in the West. Mine sites occasionally feature forge bellows and the remains of simple hand-cranked blowers, and they often feature the remains of canvas windsails and ventilation tubing.

Chapter 4: The Technology of Open Pit Mining and Blasting

1. **Peele, Robert, 1918,** p 187.

2. Ibid, p 186.

3. Ibid.

4. **Barab, J., 1927,** p 43.

5. **Young, George, 1946,** p 95.

6. **Barab, J., 1927,** p 49, 54.
Ford, Brent, Mining Engineer, Superintendent of Drilling and Blasting at Rawhide Mine, NV, 1993.
Hearon, H. H., 1959.
Peele, Robert, 1918, p 181.
Young, George, 1946, p 150.

7. Ibid, p 89.

8. **Barab, J., 1927,** p 49, 81.
E.I. DuPont de Nemours & Co., 1918, p 92.

9. **Peele, Robert, 1918,** p 327.
Young, George, 1946, p 88.

10. **Barab, J., 1927,** p 49, 41.
E.I. DuPont de Nemours & Co., 1918, p 92.
Johnson, Walter, Mining Engineer 1930s - 1960s.
Peele, Robert, 1918, p 327.
Young, George, 1946, p 89.

11. **Peele, Robert, 1918,** p 188.

12. **Argall, George Jr., 1952.**
Ford, Brent, Mining Engineer, Superintendent of Drilling and Blasting at Rawhide Mine, NV, 1993.

13. **Barab, J., 1927,** p 49, 51.
E.I. DuPont de Nemours & Co., 1917a, p 25.
McFarlane, George, 1907.

14. Ibid.

15. **Barab, J., 1927,** p 49, 51.
E.I. DuPont de Nemours & Co., 1917a, p 25.

16. **Johnson, Walter, Mining Engineer 1930s - 1960s.**

17. **Barab, J, 1927,** p 49, 81.

18. Ibid, p 74.

19. **Young, George, 1946,** p 137.

20. **E.I. DuPont de Nemours & Co., 1920,** p 63.
Young, George, 1946, p 137.

21. **Peele, Robert, 1918,** p 192.
Young, George, 1946, p 129.

22. **Peele, Robert, 1918,** p 192.

23. **Marvin, Theodore, Editor, 1927b.**

24. **Young, George, 1946,** p 149.

25. **Barab, J., 1927,** p 51, 59
Peele, Robert, 1918, p 193.

26. Ibid, p 193.

Chapter 5: Quarrying and Blasting

1. **Burgoyne, Sir John, 1849,** p 3.
Peele, Robert, 1918, p 199.

2. **Gillette,Halbert P., 1907,** p 195.
Peele, Robert, 1918, p 200.

3. **Peele, Robert, 1918,** p 201.

4. **Kirk, Aurthur, 1891,** p 14.

5. **E.I. DuPont de Nemours & Co., 1920,** p 110.
Gillette,Halbert P., 1907, p 124.
Peele, Robert, 1918, p 170.

6. **Kirk, Aurthur, 1891,** p 14.
Rothwell, Richard, Editor, 1892.

7. **Gillette,Halbert P., 1907,** p 205.
Rothwell, Richard, Editor, 1892.

8. Ibid.

9. **E.I. DuPont de Nemours & Co., 1920,** p 110.

10. **Gillette,Halbert P., 1907,** p 197.
Peele, Robert, 1918, p 200.

11. **Burgoyne, Sir John, 1849,** p 23.
E.I. DuPont de Nemours & Co., 1920, p 111.

BIBLIOGRAPHY

A New Safety Explosive *Engineering and Mining Journal* March 29, 1902.

Aetna Powder Co., 1900
Aetna Powder Co. Aetna Powder Co., Chicago, IL.

Aetna Powder Co., 1907
Aetna Dynamite Aetna Powder Co., Chicago, IL.

Aetna Powder Co., 1912
Aetna Dynamite - Blasting Stumps Aetna Powder Co., Chicago, IL.

Allen, A. W., Editor, 1930
Use of "Wooden Powder" In-Advisable *Engineering and Mining Journal* July 10, 1930.

American Well Works, 1905
The American Well Works American Well Works, Dallas TX.

Anderson, A E, 1912
Packing, Transportation, and Storage of Explosives *Mining Science* June 20, 1912.

Andre, George, 1878
Rock Blasting E. & F. N. Spon, London.

Argall, George Jr., 1952
Rotary Blast-hole Drilling in Igneous Rock *Mining World* Sept. 1952.

Atlas Powder Co., 1933

Explosives and Other Products Atlas Powder Co., Wilmington, DE.

Atlas Powder Co., 1946
Catalogue No. 10 Atlas Powder Co., Wilmington, DE.

Atlas Powder Co., 1960
This is Atlas Powder Atlas Powder Co., Wilmington, DE.

Atlas Powder Co., 1962
The Atlas Family Atlas Powder Co., Wilmington, DE.

Atlantic Dynamite Co., 1880
Giant Powder Atlantic Dynamite Co., New York, NY.

Bancroft, George, 1907
Notes on the Use and Handling of Dynamite *Mining Reporter* Sept. 5, 1907.

Bancroft, George, 1908
Editorial *Mining Reporter* July 23, 1908.

Bancroft, Peter, and Weller, Sam, 1993
Cornwall's Famous Mines *Mineralogical Record* Aug./July, 1993.

Bandmann, Nielsen & Co., 1884
Dynamite, Gelatine Dynamite, and Judson Powder Bandmann, Nielsen & Co., San Francisco, CA.

Banks, George, 1929
Dynamite Days *Explosives Engineer* Nov. 1929.

Barab, J., 1927
Modern Blasting in Quarries and Open Pits Hercules Powder Co., Wilmington, DE.

Barab, J., 1933
Dynamite, the Fore Runner of Progress *Explosives Engineer* May, 1933.

Blasting Bulletin Board *Details of Practical Mining* McGraw-Hill Book Co., New York, NY 1916.

Bohannan, Mark, 1991
Black Blasting Powder *Mining Artifact Collector* Winter 1991.

Boies, Henry, 1873
Improvement in Packages of Powder For Blasting U.S. Pat. Office, Letters Patent No. 144434, Government Printing Office, Washington, DC.

Bragg, Melody, 1990
Windows to the Past Gem Publications, Glen Jean, WV.

Brubaker, Lynn, 1962
The Production of Anthracite Coal Mining Explosives In America, 1818 - 1920 University of Deleware.

Burgoyne, Sir John, 1849
Blasting and Quarrying of Stone John Wheale, London.

Burning Empty Dynamite Cases *Colliery Engineer* Jan. 1915.

Cabinet of Miscellaneous Info *Mining Science* Dec. 26, 1926.

California Cap Company, 1910
Pointers About Blasting California Cap Co., Oakland, CA.

California Powder Works, 1878
Hercules vs Giant California Powder Works, San Francisco, CA.

California Powder Works, 1879
Superiority of Hercules Powder Joseph Winterburn & Sons, San Francisco, CA.

Canadian Indistries, Ltd., 1940
Blasters' Handbook Canadian Indistries, Ltd, Montreal, Quebec.

Charging Blast Holes *Mines and Minerals* Aug. 1911.

Chellson, H. C., 1936
From Gunpowder to Modern Dynamite *Engineering and Mining Journal* May 1936.

Cleveland Pneumatic Tool Co., ca 1906
Bulletin No.10: Cleveland Air Hammer Drills Cleveland Pneumatic Tool Co., Cleveland, OH.

Cleveland Rock Drill Co., 1928
Bulletin No.11: Cleveland A1 Hammer Drill Cleveland Rock Drill Co., Cleveland, OH.

Cleveland Rock Drill Co., ca. 1920s
Bulletin No.49: Cleveland Forty-Four Rotators Cleveland Rock Drill Co., Cleveland, OH.

Cleveland Rock Drill Co., ca. 1920s
Bulletin No.35A: Pocket-in-Head Stopers Cleveland Rock Drill Co., Cleveland, OH.

Cleveland Rock Drill Co., ca. 1920s
Bulletin No.55: Cleveland D5 Drifters
Cleveland Rock Drill Co., Cleveland,
OH.

Cleveland Rock Drill Co., 1931
Driller's Handbook Cleveland Rock
Drill Co., Cleveland, OH.

Climax Powder Mfg. Co., 1890
Climax Powder For Blasting Climax
Powder Co., Emporium, PA.

Coal and Metal Miner's Pocketbook 1902,
International Textbook Co., Scranton,
PA.

Coal Mining Notes. *Mines and Minerals*
Aug. 1911.

Cohen, Stan, 1984
King Coal Pictoral Histories
Publishing Co., Charleston, WV.

Colliery Engineering Company, 1899
A Treatise on Metal Mining Colliery
Engineering Co., Scranton, PA.

Colliery Engineering Company, 1905
Coal and Metal Miners' Pocket Book Colliery
Engineering Co., Scranton, PA.

Concentrates *Mining and Scientific Press*
Dec. 24, 1904.

Connibear, William, 1924
Insurance, Mine Accidents, and
Explosives *Explosives Engineer* April,
1924.

Crampton, Frank, 1982
Deep Enough University of
Oklahoma Press.

Dannenburg, Joe, 1960-1980

*Contemporary History of Explosives in
America* ABA Publishers, Wilmington,
DE.

Davis, Tenney, 1990 [Vol. I 1941, Vol.
2 1943]
*The Chemistry of Powder and
Explosives* Angriff Press, Hollywood,
CA.

DeCamp, W.V. 1916
Sinking with Delay-Action Fuses
Details of Practical Mining McGraw-Hill
Book Co., New York, NY.

Denver Rock Drill Manufacturing Co.,
ca. 1910s
Waugh 90 Drills Denver Rock Drill
Manufacturing Co., Denver, CO.

Denver Rock Drill Manufacturing Co.,
1914
Waugh Valveless Stopers Denver Rock
Drill Manufacturing Co., Denver, CO.

Denver Rock Drill Manufacturing Co.,
1918
*Model 65: Denver Dreadnaught Drifter
Drill* Denver Rock Drill Manufacturing
Co., Denver, CO.

Denver Rock Drill Manufacturing Co.,
ca. 1920s
Model 37: Turbro Waughhammer
Denver Rock Drill Manufacturing Co.,
Denver, CO.

Denver Rock Drill Manufacturing Co.,
ca. 1910s
Model 39: Waugh Turbro Stoper
Denver Rock Drill Manufacturing Co.,
Denver, CO.

Denver Rock Drill Manufacturing Co.,
ca. 1910s

Model 60: Denver Dreadnaught Drill
Denver Rock Drill Manufacturing Co.,
Denver, CO.

Denver Rock Drill Manufacturing Co.,
1927
Drill Steel Denver Rock Drill
Manufacturing Co., Denver, CO.

Donovan, Hedley, 1976
The Miners Time-Life Books, New
York, NY.

Dooley Brothers, 1919
Miner's Drilling Machines, Tools, Mine
Supplies, Mine Tools, Etc Dooley
Brothers, Peoria, IL.

Drake, Raymond, 1983
The Last Gold Rush Pollux Press,
Victor, CO.

E.I. DuPont de Nemours & Co., 1916
Explosives For Shale and Clay Blasting
E.I. DuPont de Nemours & Co.,
Wilmington, DE.

E.I. DuPont de Nemours & Co., 1917a
DuPont Blasting Powder E.I. DuPont
de Nemours & Co., Wilmington, DE.

E.I. DuPont de Nemours & Co., 1917b
High Explosives E.I. DuPont de
Nemours & Co., Wilmington, DE.

E.I. DuPont de Nemours & Co., 1918
Blaster's Handbook E.I. DuPont de
Nemours & Co., Wilmington, DE.

E.I. DuPont de Nemours & Co., 1920
Blaster's Handbook E.I. DuPont de
Nemours & Co., Wilmington, DE.

E.I. DuPont de Nemours & Co., 1926
Blaster's Handbook E.I. DuPont de
Nemours & Co., Wilmington, DE.

E.I. DuPont de Nemours & Co., 1932
Blaster's Handbook E.I. DuPont de
Nemours & Co., Wilmington, DE.

E.I. DuPont de Nemours & Co., 1934
Blaster's Handbook E.I. DuPont de
Nemours & Co., Wilmington, DE.
E.I. DuPont de Nemours & Co., 1952
Blaster's Handbook E.I. DuPont de
Nemours & Co., Wilmington, DE.

E.I. DuPont de Nemours & Co., 1958
Blaster's Handbook E.I. DuPont de
Nemours & Co., Wilmington, DE.

E.I. DuPont de Nemours & Co., 1922
Dumorite: Bulletin No. 38 E.I. DuPont
de Nemours & Co., Wilmington, DE.

E.I. DuPont de Nemours & Co., 1937
The DuPont Co. and its Activities. E.I.
DuPont de Nemours & Co.,
Wilmington, DE.

E.I. DuPont de Nemours Powder Co.,
1908
Black Blasting Powder E.I. DuPont de
Nemours Powder Co., Wilmington,
DE.

E.I. DuPont de Nemours Powder Co.,
1909
Gelatin Dynamite E.I. DuPont de
Nemours Powder Co., Wilmington,
DE.

E.I. DuPont de Nemours Powder Co.,
1911a
High Explosives E.I. DuPont de
Nemours Powder Co., Wilmington,
DE.

E.I. DuPont de Nemours Powder Co., 1911b
Industrial Literature and Trade Statistics *Mining Science* Nov. 30, 1911.

Dutton, William, 1960
One Thousand Years of Explosives John C. Winston Co., Philadelphia, PA.

Dynamite, Its Advantages, Methods of Using It, Proofs of Its Superiority Over All Other Blasting Products San Francisco, CA, 1872.

Eaton, Lucien, 1924
Mining Methods at the Cliffs Shaft Mine *Explosives Engineer* Feb. 1924.

Editorial *Explosives Engineer* April 1928.

Editorial: The Independent Powder Companies and Their Chance *Engineering and Mining Journal* Aug. 15, 1907.

Explosives in Anthracite Coal Mining *Explosives Engineer* Aug. 1931.

Explosives and Miscellaneous Investigations U.S. Bureau of Mines, Washington, DC ca. 1919.

Ford, Brent, Mining Engineer, Superintendent of Drilling and Blasting at Rawhide Mine, NV, 1993
Personal Interview Rawhide Mine, Rawhide, NV.

Foster, Rufus, 1913
Explosives *Colliery Engineer* Oct. 1913.

Foster, Thomas, Editor, 1896

Powder in Blasting *Metal Miner* Oct. 1896.

Foster, Thomas, Editor, 1897
Blasting in Gaseous Mines *Metal Miner* July, 1897.

Fulghum, J.T., 1944
Quarry Blasting *Explosives Engineer* March-April, 1944.

Gardner, Edward, Editor, 1927
Charging Explosives in Drift Round Holes in Some Metal Mines *Explosives Engineer* Sept. 1927.

Gardner, Edward; Howell, S P; Jones, G W, 1927
Gases From Blasting In Tunnels and Metal Mine Drifts U.S. Government. Printing Office, Washington, DC.

General Mining News: Colorado, Teller County *Mining and Scientific Press* August 24, 1912.

Gillette, Halbert, 1878
Rock Excavation: Methods and Cost Myron C. Clark Publishing Co., New York, NY.

Gonzoles, Frank, Coal Miner 1930s-1950s, 1993
Personal Interview, Aguilar, CO.

Graves, W. H., 1908
Comparisons, Classification, and Uses of High Explosives *Mining Science* Feb. 20, 1910.

Greensfelder, N. S., Editor, 1927a
Ask Me Another *Explosives Engineer* May. 1927.

Greensfelder, N. S., Editor, 1927b

Editorial *Explosives Engineer* Dec. 1927.

Gunsolis, F. H., 1910
Explosives *Mining Science* Aug. 25, 1910.

Guttman, Oscar, 1906
Blasting Charles Griffin & Co., Ltd., London.

Hall, Clarence, 1911
The Nature of Permissible Explosives *Mining Science* June 29, 1911.

Martin Hardsocg Co., 1912
Catalogue H: Spring 1912 The Martin Hardsocg Company, Pittsburgh, PA.

Hazard Powder Co., 1890
Hazard Powder Co. Hazard Powder Co., New York, NY.

Hearon, H H , 1959
Drilling and Blasting at the Jackpile Mine *Explosives Engineer* March/April 1959.

Hecla Powder Co., 1882
Hecla Powder, Safety, Storage, and Economy Hecla Powder Co., New York, NY.

Hercules Powder Co., 1882
Hercules Powder Hercules Powder Co., Cleveland, OH.

Hercules Powder Co., 1883
Hercules Powder Hercules Powder Co., Cleveland, OH.

Hercules Powder Co., 1918
Hercules Products Hercules Powder Co., Wilmington, DE.

Hercules Powder Co., 1928
Commercial Explosives Hercules Powder Co., Wilmington, DE.

Hercules Powder Co., 1929
Commercial Explosives Hercules Powder Co., Wilmington, DE.

Hercules Powder Co., 1931
Rock Tunnelling Methods Hercules Powder Co., Wilmington, DE.

Hercules Powder Co., 1935
Hercules Explosives and Blasting Supply Hercules Powder Co., Wilmington, DE.

Hercules Powder Co., 1952
Hercules Explosives and Blasting Supply Hercules Powder Co., Wilmington, DE.

Hercules Powder Co., 1959
Hercules Explosives, Blasting Agents and Supply Hercules Powder Co., Wilmington, DE.

Hill, Walter, 1875
Notes on Certain Explosive Agents John Allyn, Boston, MA.

Horay, James, Editor, 1947
Turning Back the Pages of Explosives History *Explosives Engineer* Jan./Feb. 1947.

Hoover, Herbert C., 1909
Principles of Mining McGraw-Hill Book Co., New York, NY.

Hopler, Robert, 1988
IME 75th Anniversary Institute of Makers of Explosives, Washington, DC.

Howard, Robert
Notes on Wooden Powder Kegs Personal Files of Robert Howard, Curator at Hagley Museum and Library.

Howard, Robert, 1973
Notes on Powder Kegs Personal Files of Robert Howard, Curator at Hagley Museum and Library.

Howard, Robert, 1979a
Notes on Powder Containers Personal Files of Robert Howard, Curator at Hagley Museum and Library.

Howard, Robert, 1979b
Some Notes on Papers, Kegs, and Cans Personal Files of Robert Howard, Curator at Hagley Museum and Library.

Howard, Robert, 1979c
Powder Containers *Monthly Bugle* Nov. 1979.

Howell, S P, 1931
Explosives Accidents In the Anthracite Mines of Pennsylvania, 1923-1927 U.S. Government. Printing Office, Washington, DC.

Hubbard, Douglass, 1958
Ghost Mines of Yosemite Awani Press, Fresno, CA.

Hunter, Edward, Mining Engineer 1930s - 1990s, 1994
Personal Interview 1994 Victor, CO.

Ilsley, L. C. and Hooker, A B, 1926
Electric Shot Firing In Mines, Quarries, and Tunnels U.S. Government Printing Office, Washington, DC.

Ingersoll Drill Co., 1887

Catalogue No.7: Rock Drills Ingersoll Drill Co., New York, NY.

Ingersoll-Rand Drill Co., 1906
Catalogue No.6: Rand Rock Drills Ingersoll-Rand Drill Co., New York, NY.

Ingersoll-Rand Drill Co., 1908
Catalogue No.7: Ingersoll-Seargent Rock Drills Ingersoll-Rand Drill Co., New York, NY.

Ingersoll-Rand Drill Co., 1915
Butterfly Stope Drills Ingersoll-Rand Drill Co., New York, NY.

Ingersoll-Rand Drill Co., 1916
Jackhammer Ingersoll-Rand Drill Co., New York, NY.

Ingersoll-Rand Drill Co., 1921
Leyner-Ingersoll Drifters Ingersoll-Rand Drill Co., New York, NY.

Ingersoll-Rand Drill Co., 1938
Jackbits Ingersoll-Rand Drill Co., New York, NY.

Ingersoll-Rand Drill Co., 1939
Today's Most Modern Rock Drills and A Brief History of the Rock Drill Development Ingersoll-Rand Drill Co., New York, NY.

Ingersoll-Seargent Drill Co., 1900
Catalogue No.42: Mining, Tunneling, and Quarrying Machinery Ingersoll-Seargent Drill Co., New York, NY.

International Correspondence Schools, 1906
Dynamos and Motors, Percussive and Rotary Boring, Ore Dressing and Milling

International Textbook Co., Scranton, PA.

International Correspondence Schools, 1907
Rock Boring, Blasting, Coal Cutting, Trackwork International Textbook Co., Scranton, PA.

International Textbook Company, 1899
Metal Mining Vol. I-IV International Textbook Co., Scranton, PA.

Interstate Commerce Commision, 1909
Twenty-Second Annual Report of the Interstate Commerce Commission U.S. Government. Printing Office, Washington, DC.

Institute of Makers of Explosives, 1935
Safety in the Handling and Use of Explosives Institute of Makers of Explosives, Washington, DC.

J. George Leyner Engineering Works Co., 1906
Catalog No.8: Rock Terrier J. George Leyner Engineering Works Co., Denver, CO.

Jeffrey Manufacturing Company, ca. 1906
Jeffrey Power Drills for Rock and Coal Jeffrey Manufacturing Co., Columbus, OH.

Johnson, Walter, Mining Engineer 1930s - 1960s, 1993
Personal Interview 1993, Ely, NV.

Johnson, W T, 1903
The Storage of Explosives *Pacific Coast Miner* Jan. 24, 1903.

Jones, Evan, 1935
Instructing Miners In the Use of Explosives *Explosives Engineer* May 1935.

Judson Powder Co., 1885
Judson Powder and Dynamite For Stump Blasting Judson Powder Co., Drakesville, NJ.

Keystone Driller Co., 1905
Keystone Driller Co. Keystone Driller Co., Beaver Falls, PA.

Keystone Consolidated Publishing Co., 1925
The Mining Catalog 1925: Metal-Quarry Edition. Keystone Consolidated Publishing Co.

Keystone National Powder Co., 1912
Keystone National Powder Co. Keystone National Powder Co., Emporium, PA.

Kirk, Aurthur, 1891
The Quarryman and Contractor's Guide J. W. Golder & Co., Pittsburgh, PA.

Koch, Karl, Coal Miner 1920s, 1994
Personal Correspondences March 1994, Export, PA.

Latest Market Reports *Mining and Scientific Press* June 11, 1903

Laflin & Rand Powder Co., 1900
Mining, Blasting, and Sporting Powders Laflin & Rand Powder Co., New York, NY.

Levy, Si, 1920
Modern Explosives Melborn & Bath, New York, NY.

Lewis, David, 1976
Iron and Steel In America Hagley Museum and Library, Wilmington, DE.

Looney, J. E., 1923
Handling Explosives Under Ground *Explosives Engineer* Aug. 1923.

Martin, Andy, 1991
Blasting Cap Tin Catalog Old Adit Press, Tucson, AZ.

Marvin, Theodore, Editor, 1927a
Ask Me Another *Explosives Engineer* May 1927.

Marvin, Theodore, Editor, 1927b
Blasting Method of the Utah Copper Co. *Explosives Engineer* Dec.1927.

Marvin, Theodore, Editor, 1928
Editorial *Explosives Engineer* April 1928.

Marvin, Theodore, Editor, 1931
Explosives in Anthracite Coal Mining *Explosives Engineer* Aug. 1931.

Marvin, Theodore, Editor, 1934
The Hoosac Tunnel *Explosives Engineer* Aug. 1934.

Marvin, Theodore, Editor, 1935a
Ensign - Bickford Co. *Explosives Engineer* Jan., 1935.

Marvin, Theodore, Editor, 1935b
Editorial: Powder Willows *Explosives Engineer* April, 1935.

Marvin, Theodore, Editor, 1936
One Hundred Years *Explosives Engineer* May 1936.

McAdam and Westwater, 1958
Mining Explosives Oliver and Boyd, Ltd., London.

McFarlane, George, 1907
Method of Excavating Rock in Large Masses *Engineering and Mining Journal* Aug. 3, 1907.

McGraw, James, 1901a
Explosion of Dynamite In the New York Subway *Engineering and Mining Journal* Feb. 1, 1901.

McGraw, James, 1901b
Recent Decisions Affecting the Mining Industry *Engineering and Mining Journal* Aug.17, 1901.

McGraw, James, 1901c
Recent Decisions Affecting the Mining Industry *Engineering and Mining Journal* Dec. 21, 1901.

Metal Quarry Catalogs: 1935-1936 McGraw-Hill Catalog Services, New York, NY.

Mills, W. F., 1906a
The Cabinet. *Mining Reporter* March 1, 1906.

Mills, W. F., 1906b
Editorial: Poisoning By Nitrous Fumes *Mining Reporter* March 1, 1906.

Mitchell, J. Roger, 1993
The Hammon Safety Explosives Box. *Eureka!* Winter 1993.

Mine & Smelter Supply Co., 1937
Catalog No.92: Machinery and Supplies Mine & Smelter Supply Co., Denver, CO.

Mine Air Before and After Blast Firing. *Pacific Miner* May 1909.

Mine Inspectors' Reports: File 48441 Vindicator Mine Colorado State Archives, Denver, CO.

Mining Summary: Colorado, Teller County *Mining and Scientific Press* Aug. 25, 1900.

George Mowbray, 1872
Tri-Nitro-Glycerine James T. Robinson & Son, North Adams, MA.

Munroe, Charles, and Hall, Clarence, 1909
A Primer On Explosives For Coal Miners U.S. Government Printing Office, Washington, DC.

Munroe, Charles, and Hall, Clarence, 1915
A Primer On Explosives For Metal Miners and Quarrymen U.S. Government Printing Office, Washington, DC.

Need For Ventilation *Mining and Scientific Press* Feb. 13, 1904.

Nobel's Explosives Co., Ltd., 1875
The Safey of Dynamite Robert Maclehose, Glascow.

Nobel's Explosives Co., Ltd., 1887
Settle's Patented Gelatine - Water Cartridge Nobel's Explosives Co., Ltd., Glascow.

Old Powder Said to be Cause of Accident *Engineering and Mining Journal* July 3, 1926.

Packages For Blasting Powder In Ohio. *Engineering and Mining Journal* Oct. 17, 1908.

Patents Relative to Mining and Metallurgy. *Mining and Metallurgy* Aug. 3 1901.

Peele, Robert, 1918
Mining Engineer's Handbook John Wiley & Sons, New York, NY.

Polly, Philip, Tungsten and Coal Miner, 1930s - 1940s, 1993
Personal Interview Oct. 15, 1993 Luning, NV.

Production of Explosives *Colliery Engineer* Nov. 1914.

Rand Drill Co., 1886
Illustrated Catalogue of the Rand Drill Company Rand Drill Co., New York, NY.

Repauno Chemical Co., 1883
Atlas Powder Repauno Chemical Co., Repauno, NJ.

Repauno Chemical Co., 1895
Atlas Powder Repauno Chemical Co., Repauno, NJ.

Rice, Claude T., 1916
Magazine for Storing and Thawing *Details of Practical Mining* McGraw-Hill Book Co., New York, NY.

Rickard, T. A., 1913
Missed Holes *Mining and Scientific Press* Feb. 8, 1913.

Rock Products Buyer's Directory 1945 Rock Products (no location given).

Rothwell, Richard, Editor, 1886a
Dynamite Explosions *Engineering and Mining Journal* July 10, 1886.

Rothwell, Richard, Editor, 1886b
Furnace, Mill, and Factory *Engineering and Mining Journal* Sept. 4, 1886.

Rothwell, Richard, Editor, 1890
Unfreezing Dynamite *Engineering and Mining Journal* Aug. 30, 1890.

Rothwell, Richard, Editor, 1892
Dimension Stone Quarrying *Engineering and Mining Journal* Aug. 30, 1892.

Rutledge, J. J. and Hall, Clarence, 1912
The Use of Permissible Explosives U.S. Government Printing Office, Washington, DC.

Sackett, Earl, 1977
The Ubiquitous Powder Box *Explosive and Mining Journal* Aug. 1977.

Sampson, George, 1882
Atlas Powder George Sampson, Boston, MA.

Sampson, George, 1883
Atlas Powder George Sampson, Boston, MA.

Shaffner, T. P., 1869
The Nitroglycerine Co. of New York The Nitroglycerine Co. of New York, New York, NY.

Smith, B. H., 1916
Form for Missed Hole Reports *Details of Practical Mining* Mc Graw-Hill Book Co., New York, NY.

Snelling, Walter, 1912a
Experts on the Rate of Burning Fuse *Mining Science* Feb. 15, 1912.

Snelling, Walter, 1912b
Commentary *Mining Science* Feb. 29, 1912.

Snelling, Walter, 1912c
Explosives *Mining Science* Feb. 29, 1912.

Snelling, Walter and Storm, C G, 1913
The Analysis of Black Powder and Dynamite. U.S. Government. Printing Office, Washington, DC.

Sporr, Ray, Mining Engineer and Gold Miner 1930s - 1950s, 1993
Personal Interview July 20, 1993, Delta UT.

Spur, J. R., 1918
Production and Domestic Distribution of Explosives in 1917 *Engineering and Mining Journal* July 13, 1918.

Spur, J. R., 1920
Practice of Opening Kegs of Black Powder with Wooden Tools Condemned *Engineering and Mining Journal* Oct. 30, 1920.

Spur, J. R., 1922
Most Permissibles Sold are of Ammonium Nitrate *Engineering and Mining Journal* Aug. 26, 1922.

Stack, Barbara, 1982
Handbook of Mining and Tunnelling Machinery John Wiley & Sons, New York, NY.

Stark, Jared, 1910

Accidents From the Use of Explosives in 1909 *Mines and Minerals* March 1910.

Stoek, Henry, 1897
A Chronology of Coal Mining *The Mine Bulletin* May 1897.

Stoek, Henry, 1906a
Practical Rules for Blasting Coal *Mines and Minerals* March 1906.

Stoek, Henry, 1906b
Sodium Nitrate Powder Used in Mines Today *Mines and Minerals* March 1906.

Stoek, Henry, 1906c
Powder Required for Blasting Coal. *Mines and Minerals* March 1906.

Street, Al, 1918
Negligence in Blasting Operations *Engineering and Mining Journal* Sept. 21, 1918.

Suitable Powder Magazine *Details of Practical Mining* McGraw-Hill Book Co., New York, NY, 1916.

Taylor, W. W., 1931
Good Old Black Powder *Explosives Engineer* Oct. 1931.

Testing Explosives *Mining Magazine* Feb. 1906.

Tests Show Economy In Using Large Size Dynamite *Engineering and Mining Journal* Feb. 24, 1923.

Then Came Nitroglycerine *Explosive Engineer* April 1934.

Tiffany, J. E. and Coats, A. B., 1935

Deterioration of Permissible Explosives When Stored In Coal Mines *Explosive Engineer* Feb. 1935.

The Transport of Explosives *Engineering and Mining Journal* Jan. 24, 1885.

Tratman, E. E., 1893
Note on Unfreezable Dynamite Transactions of the American Institute of Mining Engineers.

Twitty, Eric, 1997
The Electric Blasting Cap *Collectors' Mining Review* Spring, 1997

U.S. Blasting Oil Co., 1866
U.S. Blasting Oil Co. John W. Amerman Printing, New York, NY.

U.S. Department of the Interior, 1919
Explosives and Miscellaneous Investigations U.S. Government Printing Office, Washington, DC.

U.S. Patent Office
Official Gazette of the U.S. Patent Office Government Printing Office, Washington, DC.
Vol. 3 May 20, 1873
Vol. 3 June 3, 1873
Vol. 3 June 10, 1873
Vol. 4 August 5, 1873
Vol. 4 Nov. 11, 1873
Vol. 4 Dec. 9, 1873
Vol. 5 March 17, 1874
Vol. 6 Nov. 7, 1874
Vol. 9 April 4, 1876
Vol. 10 Oct. 31, 1876
Vol. 11 March 6, 1877
Vol. 11 May 22, 1877
Vol. 13 April 30, 1878
Vol. 15 May 13, 1880
Vol. 18

Vol. 30 Feb. 10, 1885
Vol. 59 May 10, 1892
Vol. 68 August 28, 1894
Vol. 70 Feb. 19, 1895
Vol. 71 May 28, 1895
Vol. 71 June 11, 1895
Vol. 120
Vol. 125 Dec. 25, 1906

Use of Tamping Bags in Blasting
Engineering and Mining Journal April 5,
1919.

VanGelder, Arthur and Schlatter,
Hugo, 1972 [1927]
*History of the Explosives Industry In
America* Arno Press, New York, NY.

Warwick, A. M., 1902
Mining Explosives *Mines and
Minerals* Sept. 1902.

Watson, Tom, Gold Miner 1930s, 1988
Personal Interview Sept. 1988, San
Jose, CA.

Wilkinson, Norman, 1984
*Lammot DuPont and the American
Explosives Industry* University Press of
Virginia.

Williams, Archibald, 1907
The Romance of Modern Mining Seeley
& Co., Ltd., London.

Willis, Bob, President of Apache
Nitrogen, 1993
Personal Interview Aug. 30, 1993,
Benson, AZ.

Wilson, Eugene, Editor, 1909
List of Permissible Explosives *Mines
and Minerals* Dec. 1909.

Wilson, Eugene, Editor, 1910a

Mining Coal with Explosives *Mines
and Minerals* Feb. 1910.

Wilson, Eugene, Editor, 1910b
Permissible Explosives *Mines and
Minerals* Feb. 1910.

Wyman, Mark, 1979
Hard Rock Epic University of
California Press, Berkeley, CA.

Young, George, 1923
Elements of Mining John Wiley &
Sons, New York, NY.

Young, George, 1946
Elements of Mining John Wiley &
Sons, New York, NY.

Young, Otis Jr., 1982 [1970]
Western Mining University of
Oklahoma Press, Norman, OK.

INDEX